*"Twenty-volume folios
will never make a revolution.
It's the little pocket pamphlets
that are to be feared."*
Voltaire

FIELD #10 NOTES

DON GILLMOR

On Oil

BIBLIOASIS
Windsor, Ontario

Copyright © Don Gillmor, 2025

All rights reserved. No part of this publication may be reproduced or transmitted in any form or by any means, electronic or mechanical, including photocopying, recording, or any information storage and retrieval system, without permission in writing from the publisher or a licence from The Canadian Copyright Licensing Agency (Access Copyright). For an Access Copyright licence visit www.accesscopyright.ca or call toll free to 1-800-893-5777.

FIRST EDITION
10 9 8 7 6 5 4 3 2 1

Library and Archives Canada Cataloguing in Publication
Title: On oil / Don Gillmor.
Names: Gillmor, Don, author.
Series: Field notes (Biblioasis)
Description: Series statement: Field notes | Includes bibliographical references.
Identifiers: Canadiana (print) 20250150999 | Canadiana (ebook) 20250151855
 ISBN 9781771966672 (softcover) | ISBN 9781771966689 (EPUB)
Subjects: LCSH: Petroleum industry and trade—Social aspects. | LCSH: Petroleum industry and trade—Economic aspects. | LCSH: Petroleum industry and trade—Environmental aspects.
Classification: LCC HD9560.5 .G55 2025 | DDC 338.2/7282—dc23

Edited by Daniel Wells
Copyedited by John Sweet
Typeset by Vanessa Stauffer
Series designed by Ingrid Paulson

Published with the generous assistance of the Canada Council for the Arts, which last year invested $153 million to bring the arts to Canadians throughout the country, and the financial support of the Government of Canada. Biblioasis also acknowledges the support of the Ontario Arts Council (OAC), an agency of the Government of Ontario, which last year funded 1,709 individual artists and 1,078 organizations in 204 communities across Ontario, for a total of $52.1 million, and the contribution of the Government of Ontario through the Ontario Book Publishing Tax Credit and Ontario Creates.

PRINTED AND BOUND IN CANADA

Contents

1. Babylon / 9

2. *Fin de Siècle* / 25

3. Mythology and Transformation / 33

4. The Battle Begins / 40

5. Through the Looking-Glass / 52

6. Shifting Sands / 69

7. Orphan in the Storm / 87

8. Oil Addiction / 104

9. The Fracking Revolution / 115

10. The Rapture / 122

    Epilogue / 129

    Sources / 135

Then I saw an angel standing in the sun, and with a loud voice he called to all the birds that fly in midheaven, "Come gather for the great supper of God, to eat the flesh of kings, the flesh of captains, the flesh of mighty men, the flesh of horses and their riders, and the flesh of all men."

Revelation 19:17–18

When it comes to oil, there are only masters, traders, and slaves.

Andrew Nikiforuk,
The Energy of Slaves

Babylon

IN 1967, JOHN Howard Pew, the eighty-five-year-old chair of Sun Oil, and Ernest Manning, premier of Alberta, attended the opening of the bitumen upgrading plant near Fort McMurray. It was part of the Great Canadian Oil Sands development, a subsidiary of Sun Oil. Both men were evangelical Christians. In 1930, Manning began preaching on a radio program, *Back to the Bible Hour,* and continued to preach as premier, encouraging Christians to live in the light of Jesus's return. Pew was on the board of the magazine *Christianity Today,* which he helped finance, though critics said the magazine was merely a "tool of the oil interests." He saw faith and oil as intertwined, and conflated both with freedom. "Without Christian freedom no freedom is possible," he said.

Manning saw both progress and redemption in the oil sands. "We should be anxious," he wrote, "for people to know about the oil which in the lamp of God's Word produces a light that shines across the darkness of this world in order that men may find their way to Jesus Christ, the one who alone can save and who can solve their problems, whatever they may be." Both men saw the oil sands as a gift from God.

I moved to Calgary four years after this baptism, in September 1971, the day the province voted in a Conservative government. It was a shift to the left from Manning's Social Credit party, a regime that had lasted thirty-six years. The city was new and white, almost Aryan-looking, defined by oil, the mountains shining to the west.

Two years later, Calgary was transformed by a holy war. On October 6, 1973, on Yom Kippur, Egyptian tanks invaded the Israeli-occupied Sinai Peninsula, and Syrian troops crossed the Golan Heights to attack Jerusalem. The United States supported Israel, and on October 16, Arab oil producers met in Kuwait and agreed to cut oil production by 25 percent in an effort to put pressure on Israel's Western allies. Production would be cut by an additional 5 percent each month until a Middle East settlement could be reached.

Within a few months, oil went from US$2.90 a barrel to almost $12, then $17 ($123 in 2024 dollars), a boon for Alberta. Calgary expanded riotously, indiscriminately. Over the next several years, between $1 billion and $2.5 billion (roughly $15 billion in 2024 dollars) worth of building permits were issued annually. In 1979, Calgary built more office space than New York or Chicago. Height restrictions were waived, setbacks ignored, permittable land uses altered, and half the city planning department laid off. Newly rich, the city was the Jed Clampett of urbanism. For thirty-six years (1935–1971) the province had been God's Country, ruled by the Social Credit party, originally led by an evangelical radio preacher, William

"Bible Bill" Aberhart, and now it was Babylon. The city expanded at the edges, suburbs rapidly built to handle all the people who flock to a boom. The downtown grew like time-lapse photography, new buildings going up, hasty and undistinguished.

Studies of boom towns tend to show a dramatic rise in real estate prices, cost of living, wealth, inequality, crime rates, and addiction. Our civic hysteria would last almost a decade.

* * *

THE MONEY WAS in oil, but you had to know someone. New to the city, I didn't know anyone, but I was looking for a summer job in the oil fields, so a friend and I drove east from Calgary in a rented Ford Maverick, looking for derricks outside Medicine Hat. "The town that was born lucky," in Rudyard Kipling's 1908 assessment, sat on a massive gas field covering 313,631 hectares, holding almost two billion cubic feet of natural gas. "All Hell for a basement," in Kipling's words. The gas was contained in sandstone from the Upper Cretaceous, some of it only 450 metres below the surface, easily accessible.

The first rig we stopped at had a sign that read "This rig has worked 0 accident-free days." A man in his twenties sat on a forty-gallon drum, head down, hand wrapped in gauze, blood staining his jeans. So, a job opening. But we didn't have any experience and the driller said he had no use for us.

We spent four days driving out to rigs in the 32°C July heat before we found one that would take us. A crew hadn't shown up for their afternoon shift and the driller looked at my friend and me, our longish hair, and said, "I guess you girls'll have to do."

The driller's name was Joe, an angry man in his mid-forties with a furrowed face that came to a point, like a villain in a children's book. I wrestled with the nine-metre drill pipe in the late afternoon heat, sliding on the slick steel floor, losing my hard hat, Joe screaming curses that disappeared in the noise of the engines. Black diesel smoke drifted toward us. The work was exhausting and confusing and there wasn't a lunch break. It was still light at 9 p.m., but dark storm clouds were moving in from the west. The air cooled within minutes and the rain arrived like an artillery attack, bouncing off the steel rig. I assumed we would go inside and wait it out, but we kept working, soaked through and chilled to the bone.

At midnight, the graveyard shift failed to arrive. The day-shift driller lived in a trailer parked near the rig and Joe sent me over to wake him. I hammered on the padded door and a man finally opened it angrily, scratching and blinking. He was in his late forties, with black hair and rockabilly sideburns. I told him we needed an extra hand and he turned away, swearing. In the kitchen, a pale girl my age leaned against the counter. She was wearing a man's shirt, her legs the colour of skim milk, her face puffy with sleep, blinking opaquely. There was an empty bottle of Black Velvet whisky on the table. The day driller came out, swearing at me, the rain, at God, at the missing graveyard shift. We worked without stopping, pulling the pipe out of the ground, changing the worn-out drill bit, then running the pipe back down. The noise of the engines and the barrage of rain bouncing off the steel made it impossible to talk, a mercy. The rain finally stopped and a lurid red sunrise brought some warmth.

* * *

MY FRIEND QUIT after a month, but I stayed. The work was dirty and dangerous, but the money was good, and it was a colourful subculture, which had a lot of appeal for an English major. And it was a way to measure myself, the young man's rite of passage, stepping into a foreign, unforgiving world. It represented some kind of freedom, though it was hard to say what kind.

Joe was recently divorced, his teenage daughter a runaway. He lived out of his truck, drunk by noon, out of his mind. He would stand over the wellhead smoking under the emphatic No Smoking! sign, telling me for the fifteenth time that people went to university to get stupider. When a safety inspector came out, the only time I ever saw one in the oil patch, he examined the cigarette butts littered across the drilling floor. "You're drilling for *gas*, you goddamn doorknob," he yelled, his face inches from Joe's. "You think there might be a connection?" He threatened to fine every man on the rig $500 and the drilling company $2,500. While he was there, I accidentally started a small grass fire while trying to fix the backup generator, which chronically misfired and was a mechanical mystery to me. Sparks flew out and ignited the dry scrub, and the crew, along with the safety inspector, ran around the prairie with wet sacks, trying to put it out. "This is really something," the inspector said. "This just about takes the blue ribbon." After he drove off, Joe lit up a Player's Plain and said, "I may have to educate that son of a bitch with a two-foot pipe wrench."

We drilled in the Palliser Triangle, a large area of southeastern Alberta that contains arid semi-desert with small cacti and rattlesnakes, as well as irrigated farmers' fields. The farmers welcomed the extra income but resented our presence, loud, unceasing, and messy. We drilled in a farmer's canola crop and Joe got in his truck and flattened everything within the lease boundary out of spite and the farmer came out with a rifle and fired three shots at the rig.

An itinerant petroleum engineer occasionally came out to the well site. He drove a Mercury Marquis and drank from a hip flask, a big man who swaggered and barked orders. Driving home through the semi-desert at midnight, Joe and I saw his car parked at an angle about ten metres off the road. The driver's door was open and the engineer was slumped in his seat. Joe stopped the truck and said, "I'd better go see if that dizzy prick is all right." He walked over and stuck his head in the door, then took the keys out of the ignition and threw them into the darkness. He walked back to the truck and said, "Passed out. Maybe we can move the rig before he wakes up."

On a graveyard shift, Joe gave me a hockey stick with a rag soaked in kerosene at the end, then lit it and told me to go to the end of the pipe where gas was escaping. I inched the flaming stick near the pipe and it exploded into flame, lighting up the prairie night with benzene, xylene, carbon dioxide, and dioxins. On the way back, I drove his truck so he could shoot at rattlesnakes with the .22 he kept on a gun rack. They were out in the morning sun, absorbing heat from the black asphalt. He drank beer, tossing the empties out the window and checking the rear-view mirror to see them smash on the road. We ate breakfast at the restaurant in the Husky gas station on the highway, and as we left, Joe yelled, "We already done our eight hours."

I was staying in the Corona Hotel in Medicine Hat, a cheap flophouse with a shared bathroom on every floor. On the sidewalk outside the Corona there was glass and blood, evidence of the previous night's fights. Rig workers only got one night off every three weeks, and they crowded all their anger and longing into that Saturday night. The hotel was filled with roughnecks and defeated middle-aged men at the end of something. A prostitute stalked the hallway, a tired woman in her forties in a faded print dress and cowboy boots. She smiled at me, missing a front tooth, and asked if I wanted a date.

In September, I told Joe I was heading back to university. "I don't imagine you'll be too hard to replace," he said.

* * *

I RETURNED TO the rigs each summer, working for different outfits, drilling south of Calgary, east near Brooks and Medicine Hat, north around Grande Prairie. The money was more than I could make anywhere else, and I was part of the provincial zeitgeist. Thirty thousand people moved to Calgary each year, drawn to oil. There was a palpable sense of power, of being at the centre of something.

My fellow roughnecks were ex-cons, failed farmers, a British alcoholic who fell into a sump pit filled with drilling mud, and a few students, one of whom quit halfway through the first shift and walked twenty kilometres back to town. There was a short, muscular derrickman who had a dent in his forehead from a pool ball thrown by a woman in a Grande Prairie bar, and an eighteen-year-old who drove a pink Cadillac and worshipped Elvis.

Forty kilometres northwest of Medicine Hat was the

British Army Training Unit Suffield, a vast training facility for British soldiers. It was big enough (2,700 square kilometres) that it was used for live firing. The rig I was working on was the first to drill on what was called the British Block. I would see British soldiers walking into town, dressed like an Elton John album cover, looking for romance. Most evenings ended with fights with locals in the bar parking lot, but they kept returning, like salmon swimming upstream, slaves to nature.

On our first graveyard shift, we heard a shell land somewhere in the distance. The sky flared briefly. The next one was closer, a third shell closer still. The driller handed me his truck keys and told me to find out where the hell they were firing from. "Tell those limey fucks they break it they bought it."

I rattled along the gravel roads as fast as I could, heading toward the buildings we had seen on our way in. I got out and told a soldier I needed to speak to his commanding officer, that it was life and death, a wartime trill moving through me. The soldier was my age, half a world from home, half asleep. The commanding officer had a clipped military moustache. I told him we were drilling for gas out there and his shells were getting very close.

"Drilling for gas?" he said. "And you are?"

"A roughneck on that rig."

"A rough *neck*," he said, emphasizing the second syllable. "I see."

He gave an order to someone who communicated it to whoever was firing the shells and I drove back to the rig as slowly as possible.

* * *

GRAVEYARD SHIFTS WERE best, the sky an expressive dome, the air finally cool. On slow drilling nights, I'd walk away from the rig's relentless noise and smoke a cigarette out on the prairie and stare at the night sky. Ovid said that all other creatures look down toward the earth, but man was given a face so that he might turn his eyes toward the stars. The morning would break, pink clouds in the east, the warmth spreading slowly.

In August, my hand was crushed by the breakout tongs on a midnight shift and I was driven to the hospital in Medicine Hat. There wasn't a doctor on duty and so the night nurse called him. He arrived cinematically drunk, staggering and slurring like Dudley Moore in *Arthur*. He told me to wash my hand in the sink and shakily poured half a bottle of Aspirin into my good hand, most of them ending up on the floor. Four days later I was on a Greyhound bus to Calgary with an infected hand that looked like an oven mitt. But my hand healed and I returned to the oil fields every summer. I bought an unreliable eight-year-old Ford Econoline van and listened to sad country songs on the radio and danced with big-haired waitresses, and every September I went back to university with a swollen bank account and tales of oil patch madness. Oil had captured the civic psyche and infected my world. It offered, more than anything else, a sense of possibility. Here was the New Rome.

* * *

DURING THE DECADE I lived in Calgary, the city was utterly transformed by oil. In 1973, the Calgary Tower (then the Husky Tower), a Jetsonian spike that sits in the middle of the city, was the tallest building in town. Its revolving restaurant was frequented by tourists and by university students on LSD who watched their steak sandwiches turn to carrion and observed the passing city in a sluggish panorama. It would be an exaggeration to say the landscape changed from one revolution to the next, but not much of one. There were usually more than a dozen cranes perched above the skyline, poised over the modest 1940s office buildings like birds of prey. Most of the sandstone structures that had defined an earlier version of the city (it was once called the Sandstone City) were gone or slated for demolition.

What went up were generic skyscrapers. Calgary was a bottom-line town, and its downtown architecture reflected that. It didn't bring beauty or enlightenment to its citizens, but rather value to its shareholders. Social critic John Ruskin wrote that buildings have an eloquence, that they speak to us of what is important. Calgary's downtown spoke of growth, growth as an aesthetic, as a moral imperative.

I lived downtown, one of the few who did, renting a basement apartment in a four-storey cinder-block building on Second Avenue run by a lunatic with a small dog. The city's spiritual heart was suburban rather than urban, partly due to its rapid growth; it came to the modern through manicured lawns and new cars—it had more cars than people. During the week, the core had an infectious energy, but on Sundays it was empty, so empty that a low-budget movie about an apocalyptic future didn't need to make any arrangements to stop traffic. There wasn't a car or pedestrian in sight. I watched them shooting and asked one of the technicians if they needed any extras. "It's the future," he said. "Everybody's dead."

The notion of whether I was a "good Calgarian," or even a Calgarian, was never formally put to me, but I felt the unstated censure of this question. In the early 1970s, half the population of Calgary weren't even born in Alberta. Yet there was a public identity. Calgary revelled in its civic character, then as now often described as "brash," a curious mixture of rebelliousness and deep conservatism. In *The Economy of Cities*, Jane Jacobs wrote that "Cities are not ordained; they are wholly existential." In Calgary, the city was existential, but the identity was ordained. In the 1970s you could still hear Texan and Oklahoman accents on the street. Calgary was more aligned with those states, its allegiances formed on a north/south axis, while it fought the enemy to the east. Pierre Trudeau's Mephistophelean face became a lasting symbol of that enemy, his National Energy Program demonized in Alberta. The vilification of Trudeau remained potent for decades, passed along to his son like a family heirloom.

* * *

IN MY UNDERGRADUATE mind I imagined a David Lynchian narrative lurking beneath Calgary's almost doctrinaire middle class—white, prosperous, righteous. This view found its epiphany in January 1995, when Earl Joudrie, an oilman of renown, visited his estranged wife, Dorothy, in her condo. They were meeting to discuss their divorce, which had been pending for five years. Dorothy was wearing a black sweater, black stretch pants, and heels, and as their conversation ended, she shot Earl in the back six times with a small-calibre pistol. He collapsed on the floor, and Dorothy went to the living room and mixed a double Seagram's VO on the rocks. She drank it, then mixed another and went to see about Earl, who was alive and conscious and lying face down in his blood. "How long is it going to take you to die?" she asked, sipping her drink.

Earl offered her a deal: If she called emergency, he wouldn't press charges. Dorothy had come to the end of her plan (and her bullets) and had had several drinks and couldn't think of a clear way out of a very messy situation. She dialed 911 and Earl was saved. Dorothy was tried for attempted murder and found "not criminally responsible." There were mitigating factors: Earl had been physically abusive during the course of their marriage. Four of the bullets remained in Earl until his death in 2006.

The Joudrie shooting captured the public imagination. It was one of those crimes that reflects the civic zeitgeist, the way the Manson murders did in Los Angeles in 1969, or the assault on the Central Park jogger in New York in 1989. The city recognized some aspect of itself. Calgary wasn't criminally responsible—wealthy, careless, and gunned on whisky.

* * *

IN BOARDROOMS IN Houston, Calgary, Kuwait, and a dozen other oil capitals, and on the floor of the New York and Chicago Mercantile Exchanges, oil was a global chess game where commodity futures were sold and bartered, oil shipped and traded. Millions of barrels lurched across the globe each day, traders hunched over streaming charts, puzzling over contracts for difference. Over the years, oil has won wars, started others, been a force for nationalism and colonization, and provided a stubborn mythology. It is the one true global religion. A glimpse of oil's reach can be seen in America's oil industry. Under Joe Biden, it produced 13.3 million barrels per day, enough to meet the US's own needs. But in 2023 it imported 8.51 million barrels per day (bpd) from dozens of countries, and exported 10.15 million bpd to 173 countries. The oil network envelops the world in a complex web of shipping and refining capacity and capability, depending on cost-effectiveness and the grade of oil. Part of this is economic; it can be cheaper to import from countries with lower labour and capital costs and fewer environmental regulations. And part of it is chemistry; the heavy, sour (high sulphur content) oil that the US was importing from Venezuela and Mexico when it still needed to import oil requires a specific kind of refinery. Some of the refineries on the Gulf coast are equipped to process that oil (along with Canadian bitumen), as opposed to the light, sweet oil that Texas produces. Refineries take years to build and are expensive—between US$5 and 15 billion. With the exception of a small North Dakota refinery that came online in 2020, no US refineries have been built since 1976. Past the economics and chemistry, there are the geopolitics. Countries (Russia, Saudi Arabia, the US) sell oil at advantageous prices to other countries to gain influence and status. It is the world's most pervasive diplomatic tool.

Canada's oil landscape is equally byzantine. Canadian pipelines tend to run south rather than east, so Ontario and Quebec get their oil from an evolving patchwork of sources that shifts depending on economics and politics.

Since 1988, eastern Canada has imported more than $500 billion in foreign oil, coming from the US, Venezuela, Saudi Arabia, Algeria, Nigeria, Norway, and others. The landscape can shift quickly. In 2012, Quebec got 92 percent of its oil from Kazakhstan, Angola, and Algeria, and just 1 percent from Alberta. Five years later, it was getting 44 percent of its oil from Alberta, the result of Enbridge's Line 9 pipeline.

It binds us all. Oil has a pulse, it evolves and migrates, transforming cities and governments, entire countries. It fuelled economic growth and triggered recessions and gave us the romance of the open road. But at its source, in Texas and Oklahoma and Louisiana and in camps in the Arctic, and outside Medicine Hat, it was men trudging onto the drilling floor, labouring in the heat or cold amid a symphony of engine noise, wrestling with drill pipe, spinning chains, tongs and slips, the kelly hose bobbing above them as they punched another hole in the earth. Even for us, oil remained an abstraction. I never saw it; there were no dramatic gushers, black oil spewing from the earth, coating everything. It powered our cars and homes and was used in the manufacture of a thousand products, from plastics to fertilizers to Aspirin. It powered our lives: We are Hydrocarbon Man, *Homo Oleum*. Yet it remains unseen, the ghost in the machine.

* * *

AFTER GRADUATING FROM university, I worked on an oil rig for a hundred straight days, with what was probably the oldest crew in the oil patch, weathered, gnarled men in their late sixties, one in his seventies, ancient for rig work, their lives a country and western song. Pete, the wobbly seventy-two-year-old derrickman, came home to find his wife gone, along with all the furniture, appliances, and curtains. There was a note on the floor: "Your dinner's in the oven." There was no oven. The driller was a ropy-muscled troll who had worked on the killing floor of a meat-packing plant but quit finally, saying it took too much of you, all that death. My fellow roughneck was a farmer whose modest crop had been lost to drought. He was in his late sixties, with a deeply lined face, a face that could hold a spring rain, as my grandmother might have said. Between connections he would roll a cigarette and walk to the edge of the lease and smoke and stare at the horizon.

I went up in the derrick when Pete was drunk or too hungover to climb the thirty metres onto his perch. We were south of Calgary and I was ten storeys off the ground, a view of the Rockies to the west and limitless prairie to the east, farms and ranches laid out like a Mondrian painting, a glorious solitude.

With my first paycheque I bought a plane ticket to Europe, then counted the days like a convict. Four months later, I was sitting on a beach on the Greek island of Crete, blobs of sticky oil dotting the sand around me. A tanker carrying crude oil from Libya had run aground off the south coast of Crete and here was the residue. Only two months earlier, the *Amoco Cadiz* had split apart off the Brittany coast in France, spilling 230,000 tonnes of oil, at that point the largest spill in history. Twenty thousand birds were killed and millions of sea creatures. Two months after the spill, six thousand French soldiers were still cleaning up the coastline.

The 1970s was a banner decade for oil tanker spills. More happened in that decade than any decade before or since. It was peak spill, with an alarming 788 of them (by contrast, the 2010s saw 63 spills). Millions of tonnes spilled into the seas. The world was awash in oil.

Fin de Siècle

IN THE 1970S, the environmental movement was still in its infancy. Rachel Carson's *Silent Spring* had been published in 1962, warning of the indiscriminate use of pesticides. Greenpeace was founded in 1971 to protest US nuclear testing on Amchitka Island. Environmental groups targeted whales and rainforests, but there was little organized protest around climate change. There were occasional warnings from scientists about the long-term impact of fossil fuels. Gordon MacDonald, a geophysicist, had warned the US government as early as 1965, and in 1978 he appeared on *The MacNeil/Lehrer Report*, telling us the earth was warming. He predicted that humans would eventually create an environment that was a weapon of mass destruction. MacDonald was part of an elite science team that advised US president Jimmy Carter, and in the summer of 1978 they met in Woods Hole, Massachusetts, to look at climate models. Their consensus was that the earth would warm by 3°C in the next century, but their report wasn't publicized, and the generous timeline deflected any immediate concern for politicians.

In Alberta, there were complaints about oil and gas development, but they were mostly local and unorganized—farmers and ranchers who worried their land or water was being contaminated, their voices lost in the din. The seventies in Calgary was a drunken party where the bar never closed.

Warnings about climate change go back as far as 1824, when Joseph Fourier published his theories about the earth's atmosphere. He studied equations relating to heat transfer and calculated the energy being supplied by the sun. By his calculations, the earth should have been colder than it was. He decided there must be something in the atmosphere that prevented heat from escaping, and that over time the amount of heat held by the atmosphere could conceivably change. "The establishment and progress of human society," he wrote, "and the action of natural powers, may, in extensive regions, produce remarkable changes in the state of the surface, the distribution of the waters, and the great movements of the air. Such effects, in the course of some centuries, must produce variations in the mean temperature for such places."

In 1896, Swedish physicist Svante Arrhenius created what may have been the first model for climate change. The math was complicated, and it took more than ten thousand calculations to predict how much heat would be trapped if CO_2 levels changed. The work took up most of 1895. The following year, he produced a carefully worked-out prediction: If CO_2 levels doubled, it would raise the world's temperature by 5 to 6 degrees Celsius, a prediction that is proving to be disturbingly accurate.

Sixteen years later, another distant warning. A March 1912 edition of *Popular Mechanics* stated: "The furnaces

ON OIL

of the world are now burning about 2,000,000,000 tons of coal a year. When this is burned, uniting with oxygen, it adds about 7,000,000,000 tons of carbon dioxide to the atmosphere yearly. This tends to make the air a more effective blanket for the earth and raise its temperature. The effect may be considerable in a few centuries."

By the 1950s, there were more furnaces, most of them burning oil or gas. By the sixties, a few scientists, including Gordon MacDonald, examined the possible climate outcomes, their timelines now shortened, the consequences no longer centuries away. In 1968, MacDonald published as essay, "How to Wreck the Environment: Anthropogenic Extinction of Life on Earth," but his doomsday scenario failed to capture the public imagination.

If the public wasn't listening, the CIA was. In August 1974, it produced a study on "climatological research as it pertains to intelligence problems." It warned of dramatic changes in weather. "The climate change began in 1960," the report said, "but no one, including the climatologists, recognized it." This specific start date wasn't accurate and likely referred to crop failures in India and the Soviet Union, the most recent evidence. The CIA was focused on the human toll of climate change, specifically the effect on the US. There would be mass migration as parts of the world became uninhabitable, which would lead to political unrest and, ultimately, war.

The report was made public in 1977, and concluded with the hope "that the current crisis is severe enough and close enough to home to encourage the interest and planning required to deal with these long-range issues before the problems get too much worse."

The CIA assumed, with uncharacteristic naïveté, that

climate experts and energy producers would work together to deal with this looming existential threat. Instead, they became adversaries.

The oil industry was doing its own research on climate change. At a petroleum conference at Columbia University in 1959 called "Energy and Man," scientist Edward Teller (one of the inventors of the hydrogen bomb) warned that if nations continued to use oil, eventually the ice caps would start melting and the level of the oceans would begin to rise. The American Petroleum Institute (API), which represents oil and gas interests in the US, formed a climate task force in 1979, and the following year Dr J.A. Laurman described the complexities of climate science and concluded with this grim scenario: "At a 3% per annum growth rate of CO_2, a 2.5 C rise brings world economic growth to a halt in about 2025."

An internal Exxon memo from 1979 said an increase in CO_2 concentration will "cause a warming of the Earth's surface" and the "present trend of fossil fuel consumption will cause dramatic environmental effects before the year 2050. The potential problem is great and urgent."

The problem *was* both great and urgent, but it was almost entirely unglimpsed by the public. Like the tobacco industry, oil understood the dangers of its product and started to work to suppress any science that exposed those dangers. The API disbanded the climate task force in 1983, then turned its attention to actively undermining climate science. The environmental unit evolved into a lobbying group.

Had we known of the dangers back then, what would we have done? Forty-five years after the Exxon memo, armed with a thousand dire facts and fatal predictions, faced with rising seas and the red, apocalyptic skies of wildfires, we have done relatively little. We would have done less back then. But perhaps the debate might have started sooner, and more progress could have been made.

Part of the problem with predictions of doom is how the human brain processes distant threats, which is not very effectively. Our brains are designed for short-term survival, but we don't process long-term threats, neither as individuals nor as a society. We are hard-wired to ignore climate change. Concern for the future of the planet was mirrored in our boomer psyches. Live for the moment was the hipster philosophy. We heard warnings about alcohol, drugs, sex, and the damage Jimi Hendrix was doing to our eardrums. But we were young, and our hearts were strong.

* * *

IN THE 1970s, Alberta's oil sands were in their infancy, still a few decades away from becoming the poster child for greenhouse gas emissions. They were viewed as an expensive folly by many in the conventional oil industry. The first large-scale oil sands mine and bitumen upgrader only started in 1967. Syncrude, the largest oil sands operator at the time, didn't start production until 1978. Mining bitumen was expensive and logistically challenging, but the numbers were compelling: 1.8 trillion barrels of bitumen (though only 168 billion were recoverable using existing technology under current economic conditions). Enough to power civilization until it ends.

* * *

IN THE SEVENTIES, there were no unions in the oil fields and little meaningful environmental oversight. There had been accidents—collapsing derricks, blow-ins, hydrochloric acid from fracking leaking into waterways, dozens of fingers lost to the spinning chain—though for the most part it all happened offstage, in remote areas. But in 1982, Amoco Canada had a major sour gas well blow-in near Lodgepole, a hamlet in the middle of the province, that caught the media's attention. For sixty-eight days, it released 200 million cubic feet of poisonous hydrogen sulphide sour gas per day, along with other toxins.

Hydrogen sulphide gas has the distinctive smell of rotten eggs, is carcinogenic, and is lethal in significant doses. It is heavier than air, and hugs the ground rather than rising. It rapidly attacks the organs, and in high doses can induce convulsions and coma and, finally, death. The gas killed two people, sixteen were hospitalized, and thousands became sick. The gas caught fire, producing a plume of flame that shot up hundreds of feet, releasing other toxins into central Alberta then drifting into Saskatchewan on the westerly winds. People and livestock were evacuated.

Both Amoco and the government denied any responsibility and downplayed the environmental impact and the effect on human and animal health. In that unsettling lacuna, the Pembina Area Sour Gas Exposures Committee was formed, which forced a public inquiry. Out of that came a series of recommendations regarding drilling activity, many of which were adopted (though some have since been rescinded). The initial committee morphed into the Pembina Institute for Appropriate Development, and became an early environmental watchdog for the Alberta oil and gas industry.

The year of the Lodgepole crisis was also the year the boom ended. It had lasted almost a decade. There was a broad civic sense that it couldn't last, that a Calvinist judgment was coming. All that reckless wealth. Oil was $35 a barrel ($122 adjusted for inflation), a stratospheric price, and house prices moved in lockstep. The wealth, of which I had no real share, was leveraged, untrustworthy. That year I went to a Halloween party at a house south of town with my statuesque girlfriend. We went as Rock Hudson and Doris Day in the movie *Pillow Talk*; I was Doris, she was Rock. The underlying *fin de siècle* feeling was heightened by a fierce blizzard that arrived near midnight. We left at 2 a.m. and drove north then skidded off the road. We were pulled out of the snowbank by kind locals in a truck with a winch. Standing in that blizzard at 2:30 a.m. wearing a baby doll nightgown and a blond wig, my pink princess telephone dangling, my ancient Volvo slowly pulled out of the snow, I sensed it was the logical end of something.

The price of oil collapsed, and many of those who had come for the boom went back east, to the Maritimes or Ontario. I moved east as well. For years, Calgary's vacancy rate had approached zero, and now landlords were offering free trips to Las Vegas to anyone who signed a one-year lease. There was 25 million square feet of vacant downtown office space. Those high-rises that had looked like progress now looked like abandoned toys. By March 1984, there were 250 house foreclosures per month. The classified pages of the *Calgary Herald* were filled with houses, trucks, and furniture for sale. Bumper stickers appeared—"Please, Lord, send me another oil boom and I promise not to piss this one away." But we did, of course;

such is the nature of booms. The city became more introspective, more humane. Oil was still an economic force, but it was a background hum, a distant factory.

With the Lodgepole blow-in, the essential duality of oil was established in Alberta: at once economic engine and destroyer of worlds. This was the beginning of a war that was complex and Sisyphean and wouldn't gain much traction for several years, a war fought with science, celebrity, misinformation, and political cudgels. Sometimes the war was fought within ourselves, at once hostage and junior partner. It rages still.

Mythology and Transformation

THE WORLD'S FIRST oil rig was constructed in 1847 near Baku, Azerbaijan. In 1851, oil was discovered near Sarnia, Ontario, though ultimately it didn't yield much. The discovery of oil near Titusville, Pennsylvania, in 1859 signalled the start of an energy revolution, though, and its first gusher, which produced 3,000 bpd, arrived two years later. It ignited, killing nineteen people, and in its first two years oil prices fluctuated from ten dollars a barrel to ten cents, a harbinger of oil's volatility, both literally and economically. In 1870, John D. Rockefeller's Standard Oil was founded and quickly grew into one of the largest corporations in the world. By the end of the century it controlled 90 percent of US oil refining. Oil was hailed as the "new light," replacing whale oil, and it transformed the American economy.

In 1901, Pattillo Higgins was one of the first to hit oil at Spindletop, outside Beaumont, Texas. A year later there were 1,500 oil companies operating in the state, a handful of which survived, among them the Gulf Oil Corporation, the Texas Company (eventually Texaco), and the Sun Oil Company (Sunoco). Beaumont grew from 8,500 people to 50,000 within a year. Spindletop provided 40 percent of America's oil that year and established Texas as America's dominant producer, a state that would become defined by oil.

As a teenager, Pattillo Higgins already had a reputation for drinking and fighting, and at the age of seventeen he shot and killed a Texas deputy sheriff. Higgins was shot in the arm, which was amputated below the elbow. He was charged with murder but acquitted, then found Jesus at a Baptist revival meeting. He carried a bible and claimed to be in touch with the divine, becoming known as "the Prophet of Spindletop." He used his oil money to help build a spiritual kingdom that anticipated the return of Christ. He was the beginning of what became an enduring trend, Texas wildcatters who leaned toward an evangelical brand of Christianity.

John D. Rockefeller was a devout Baptist as well, and he believed that oil and gas "were the bountiful gifts of the great Creator" and should be used toward his kingdom on earth, though he also saw the sin that was created in its wake—the brothels and bars and immorality that accompanied oil booms—"the satanic new world bequeathed by the oil boom," as his biographer noted. For all his ruthlessness and predatory tendencies, Rockefeller was a great philanthropist. But his causes tended toward liberalism, education ($35 million to the University of Chicago alone), and medical research. Texas oilmen went in another direction. "They shored up their core principles, which were intensely evangelical," wrote historian Darren Dochuk, author of *Anointed with Oil: How Christi-*

anity and Crude Made Modern America. "They emphasized the primacy of soul-winning evangelism over social restructuring in anticipation of Christ's impending return."

Through coincidence or divine providence, oil is often found in places where evangelical Protestants are already rooted—Texas, Oklahoma, Alberta. Historically, there has been a strong streak of evangelism among oilmen, and their faith has proved to be unshakable. In 1937, in New London, Texas, a school built for the children of oil workers was destroyed by a gas explosion, killing three hundred children and staff. It was written off as God's will, though one East Texan noted, "When you strike oil, you let loose Hades." Some of the Texas oilmen held to premillennial beliefs, that the oil would eventually run out and End Times would follow.

The Spindletop find in 1901 did more than transform the American economy; it gave birth to a mythology. The image of Texas wildcatters punching holes in the desert defined the industry as one that prized individualism, independence, and perseverance. It stood in for a powerful symbol of American capitalism, one where fortunes could be made overnight by the hard-working or the lucky. Oil transformed Dallas, Midland, Houston, and other cities and towns. It fuelled the American Dream. In Texas, the industry adopted the dress of America's staunchest myth, the cowboy. Oilmen wore hats and boots and western belt buckles, a mythology that was mirrored in Alberta, that sense of individual freedom and open skies. Oil as liberating force.

The idea that oil was the purest form of capitalism, a gift from God, and the underpinning of American freedom would resurface with a vengeance more than a century later.

* * *

ON MAY 14, 1914, the Dingman No. 1 Discovery Well blew in at Turner Valley, about sixty-five kilometres southwest of Calgary. By July, there were five hundred newly incorporated oil companies in Calgary, few with any equipment or expertise. Some had jars of sewing machine oil in their window labelled "Genuine Dingman Gold!"

Calgary already had an outsized sense of itself and its destiny; it had grown from 5,000 people in 1905 to 75,000 five years later, mostly in the hope of oil salvation. Thomas Mawson, a British planner and architect, came through Calgary on his way to Banff, and recognized an opportunity. He gave a speech to the Canadian Club of Calgary titled "The City on the Plain and How to Make It Beautiful." The biblical reference was prescient. In Genesis, God destroyed the cities on the plain, among them Sodom and Gomorrah, as well as their less famous neighbours Admah and Zeboiim. And much of the Calgary that was being built, the lovely sandstone buildings, would eventually be razed during the Gomorrah-like 1970s to make way for the new.

Mawson talked about Haussmann's Paris and the City Beautiful movement. He cited Daniel Burnham, who had recently published his grand plan for Chicago, and said, "Make no little plans; they have no magic to stir men's blood." Calgary's city council voted unanimously to hire Mawson to draw up an ambitious plan for their city. He delivered a series of watercolours that showed a huge central plaza, sculpted gardens, and beautiful École des Beaux-Arts buildings set on a grid. The Fourth Street bridge over the Bow River was modelled after the Pont Alexandre in Paris. Two oversized railway stations faced one another across a dramatic plaza.

It was while this proposal was being debated by city council that the Dingman blew in, ushering in what would be an era of untold wealth. But the Dingman proved to be a false spring; it didn't produce much oil, and a few months later war was declared in Europe. Mawson's grand plan was never implemented.

* * *

THE FIRST WORLD War was a turning point for oil. Just before the war began, Winston Churchill, at the time First Lord of the Admiralty, presciently secured 51 percent of the Anglo-Persian Oil Company (now British Petroleum), which was fundamental to the Allied military strategy. The conflict began as a nineteenth-century war, with men, my grandfather among them, charging the enemy on horses, sabres drawn. It ended with airplanes and tanks and unimaginable carnage, a twentieth-century war fuelled by oil.

The geopolitics of oil began after the war, precipitating a scramble for reserves in Persia and Mesopotamia, back when Iran and Iraq still had biblical names. Oil now determined military power, and control of supplies was key to national security. Iraqi oil was divided up among the victors: Britain, the US, and France. The Lausanne Treaty of 1923 set the tone. British diplomat Nevile Henderson wrote, "The end of the Lausanne Conference is indeed a sordid anti-climax: squabbling over money and rights of capitalists, with an America fouled by oil in the background..."

Oil was a key factor in both world wars, and in each case Germany was disastrously outflanked, hastening its defeat. The Oil Wars were fought on several fronts—military, diplomatic, and corporate—but they never ended. In the coming decades, oil would keep getting divided up among dominant powers (the 2003 invasion of Iraq being the most notable example). Sometimes those powers were countries, other times corporations, with the line between them blurring.

* * *

ALBERTA FINALLY GOT its deliverance in February 1947, when the Leduc No. 1 well blew in south of Edmonton. As Spindletop transformed Texas, Leduc changed Alberta. Imperial Oil had drilled 133 consecutive dry holes before Leduc came in, but that discovery ushered in the first real boom, opening up a rich new field. In 1946, Alberta had 543 producing wells, with annual production of 6.7 million barrels; ten years later, there were 7,400 wells producing 144 million barrels. The seventies saw an unprecedented increase in producing wells, and oil sands production came online, though its initial output was modest (30,000 bpd).

The premier throughout the 1970s boom was Peter Lougheed, a handsome, charismatic man who clearly laid out the relationship between the province and its resources. "Well, it was obvious that the oil sands were owned by the people of Alberta," he said in a 2011 interview, a year before his death. "We constantly and consistently made sure the industry understood that the Government of Alberta was the owner, and we weren't just there in a supervisory or regulatory way. We were there as owner and that was the constant stress of our time in government."

The oil royalty rate under the Social Credit regime had been 16.7 percent, but upon taking office in 1971, Lougheed immediately raised it to 25 percent. After the 1973 OPEC crisis, when oil prices quadrupled within months, royalties rose to 35 percent, and as high as 65 percent in some cases. When Syncrude proposed its first oil sands operation, they were looking for a break on royalties, threatening to walk away from the project if their demands weren't met. Lougheed called their bluff and held firm, insisting on a 50 percent royalty on net profits. Syncrude caved, and the Alberta government ended up buying a 10 percent stake in the operation.

Lougheed made it clear that the people of the province were the landlords and the oil companies were the tenants. But this relationship would evolve, and with the Ralph Klein administration in 1992 it was effectively reversed, with oil companies becoming feudal rulers, the citizens of Alberta reduced to wealthy peasants (its average annual income of $77,700 remains the country's highest). The mythology remained intact, though; oil was Alberta's central intoxicant, key to the shit-kicking, populist image it exported with such care.

The Battle Begins

IN 1980, RONALD Reagan became president of the United States and appointed James Watt, a determined anti-environmentalist, as secretary of the Department of the Interior. Watt described environmentalists as "a leftwing cult dedicated to bringing down the type of government I believe in," and refused to meet with them. Watt was a devout Christian who believed the End Times were near. "I do not know," he said to Congress in 1981, "how many future generations we can count on before the Lord returns." In the meantime, there wasn't much point in preserving the environment. Reagan concurred, telling television evangelist Jim Bakker, "We may be the generation that sees Armageddon."

Anne Gorsuch, a lawyer who was scornful of climate science (and whose son Neil sits on the Supreme Court), was given the role of Administrator of the Environmental Protection Agency, and cut the EPA budget by 22 percent and staff by almost 30 percent. Enforcement declined by 79 percent during her first year. She hired people from the industries the EPA was supposed to be regulating, tried to weaken pollution standards, and facilitated the use of restricted-use pesticides. She resigned in 1983 amid scandal, accused of withholding $6 million in federal funds to clean up toxic waste near Los Angeles because she was afraid it might help Democrat Jerry Brown's Senate campaign. Her most lasting legacy may have been to solidify the political battle lines around oil and the environment: If you were a Republican, you were pro-development and, if not anti-environment, at least anti-environmentalist. The debate over climate change and the environment has rarely been rational. It began as a corporate issue, then became a political issue and to some extent a generational issue, and finally, like so much these days, it has become a tribal issue.

* * *

REAGAN FAMOUSLY REMARKED that "trees cause more pollution than automobiles do," and under his watch, environmental standards predictably regressed. His administration issued leases for oil, gas, and coal development on tens of millions of acres of federal lands, more than any previous administration.

"It was eight lost years," said George T. Frampton, president of the Wilderness Society, "years of lost time that cannot be made up and where a lot of damage was done that may not be reparable."

A paper prepared for Reagan's successor, George H.W. Bush, echoed this sentiment.

"The Reagan Administration had a rare opportunity to reform America's flawed environmental protection programs. It has squandered this opportunity and perhaps made it impossible for a successor to design a Federal system to protect the environment at a cost the nation can afford to pay over the long term." Perhaps this paper was designed to comfort Bush. There was no point in addressing environmental issues, because his predecessor had made it impossible; Bush was off the hook. Though as a former oilman whose presidential campaign was partly financed by oil interests, it was unlikely that H.W. was going to lead the charge against Big Oil.

Coming into office, Reagan said he wanted to "deregulate America," and in his eight years in office he oversaw the deregulation of several industries. The first, in 1981, was oil. The environment suffered under this deregulation, though it could have benefited. "The great failure of this Administration," said Fred Krupp, executive director of the Environmental Defense Fund, "is that it blew the chance to streamline regulations and use marketplace incentives in an honest way to speed up environmental progress, lower regulatory costs and foster economic growth. Instead, the Administration abandoned any effort at progress and just accepted the weakest possible rules to protect the environment."

In 1989, a few months after Reagan left office, the *Exxon Valdez* spilled 260,000 barrels of oil into Prince William Sound, Alaska. It was the largest spill in US waters and affected 2,100 kilometres of coastline. The effect on the ecosystem was devastating: 250,000 dead seabirds, 3,000 otters, 250 bald eagles, and 22 killer whales. The footage of dead and dying oil-soaked animals was seen in news reports and left an impression that other spills hadn't. The powerful visuals helped galvanize public opinion, and the American environmental movement turned its attention to oil.

* * *

IN 1989, RALPH Klein became Alberta's environment minister. He shared some of the same qualities that had carried Reagan to the White House. Both had switched parties; Klein had been a Liberal before being lured to the Conservatives with the promise of a cabinet post, and Reagan voted Democrat before becoming Republican governor of California. Both were what was generously called "big picture," and both were populist politicians with an extraordinary ability to connect with voters.

Populism has been defined as "appealing to the interests or prejudices of ordinary people." It is often a vote against politics itself. An MIT paper titled "A Political Theory of Populism" defined it as "the implementation of policies receiving support from a significant faction of the population, but ultimately hurting the economic interests of the majority." Klein's administration was a vivid illustration of this definition.

Ernest Manning, who had been Alberta's premier from 1943 to 1968, was also a populist, but his was a different brand. His Christian principles emphasized self-reliance and the importance of family and faith, but he was a shrewd manager of the oil revenue that flowed in and used the money to fund hospitals, education, and social services. His honesty and integrity were above reproach. What had changed over the decades was the nature of populism. Manning was guided by Christian principles, while the populism of Reagan and Klein was either crafted by corporate interests (Reagan) or exploited by them (Klein), a triumph of image over principle.

I spent a few days with Klein in 1989, reporting on his transition from municipal to provincial politics. He had been a three-term mayor of Calgary, the most popular mayor in the city's history (on his third term he received 94 percent of the vote, his campaign slogan long reduced simply to "Ralph"). He was an affable man, visibly hungover when I met him at City Hall, still twelve years away from a tearful public admission that he had a drinking problem. This wasn't news to Albertans; in Calgary, his drinking had become a perverse form of civic pride, just one more connection to what the media often called "the common man." Klein was a common man among common men, an *Über* common *mensch*. Who knows what principles were buried beneath that plebeian facade? When I spoke to Calgary's chief commissioner, he said, "The Klein administration? There is no Klein administration." Early on, Klein had made a deal: Let the commissioners run the city and they would give him the money he needed for his public relations job. Klein was populist politics distilled to its essence: an enviable image, an eternal campaign.

We were going to the University of Calgary, where he was delivering a lecture on civics and how to own your mistakes, an area he excelled in. He insisted on driving and got lost, coming to a stop in a residential cul-de-sac, slumped over the wheel, saying, "I think I have the flu." The day ended fifteen hours later, at 1 a.m., Klein drinking happily with the police chief ("best damn mayor the city ever had").

Before he was mayor, Klein had been a television reporter, a portly, rumpled presence who decried the secrecy of land deals and what developers were doing to the city. As mayor, he reversed this position, telling the Rotary Club, "In our efforts to make the commercial core a more human place we must be careful to avoid the trap of putting sunlight ahead of commerce: sunlight doesn't turn the wheels in our factory."

One of Klein's legacies as mayor was indiscriminate development. The city was growing at a dizzying pace, and Klein stepped aside and let it sprawl. As an environmental design professor lamented at the time, anyone could build anything anywhere. Klein carried that philosophy into the premier's office. It was a surprise to many when he was elected premier, though perhaps it shouldn't have been. He was a shrewd politician who understood modern power. It is a testament to his political skills that throughout his career he repeatedly capitulated to corporate interests but retained the image of someone who was not only fighting for the Little Guy but was, in fact, the Little Guy.

Klein was elected premier in 1992, months after the Earth Summit in Rio de Janeiro, the first major international symposium on the environment. It was under Klein's administration that oil solidified its hold on the province, the line between oil and government blurring then effectively erased. Oil companies became the landlords, and in Klein they found a willing, impressionable manager, someone who could go door to door and placate the new tenants. This is what political scientists call "regulatory capture," the process whereby a politician or government is co-opted to serve the economic or political goals of a special interest group at the expense of the public good.

Klein's oil sands policy derived from the National Task Force on Oil Sands Strategies, created by the Alberta Chamber of Resources (ACR), an industry association that included oil producers, oil and gas servicing companies, and pipeline operators. The president of the ACR was Eric Newell, who was also president of Syncrude, the largest oil sands producer. The task force had 57 members, 45 of them from industry, with all six committee chairs filled by industry, including two from Syncrude. It was heralded as a government–industry coalition, but it was run by industry working in its own interests rather than the interests of the public. One of its recommendations was to drop the oil royalty rate from 25 percent to 1 percent, which Klein did. As Gillian Steward pointed out in her analysis of oil sands policy, "No longer would government be overseers and financial partners—they would be mere facilitators, removing obstacles on the road to development so the industry could forge ahead on its own terms."

* * *

GLOBALLY, OIL HAS been the most successful industry when it comes to regulatory capture. Political scientist Terry Lynn Karl coined the term *petrostate*, and in her seminal study of oil-producing nations in the 1970s, she concluded that while oil brought wealth, it tended to erode democracy and contribute to inequality. Authoritarian governments in oil-producing countries often neglect health, education, and social services, and put a disproportionate amount of money into fuel subsidies. There is less diversification, and both labour and capital are swallowed up by oil.

It wasn't just authoritarian governments, though. Alaska qualifies as a petrostate, ranked as the US state with the highest percentage of economic contributions by the oil and gas industry. A 2017 study noted that repealing state taxes had contributed to "a decline in government responsiveness, as measured by whether changes in citizen preferences cause changes in government policies... While we do not expect the state of Alaska to succumb to dictatorship, the loss of government representation as a consequence of natural resource wealth is troubling."

Low taxes are one of the hallmarks of petrostates, a way to placate its citizens (Texas, Wyoming, and Alaska have no tax, Saudi Arabia has no personal tax, the United Arab Emirates has the lowest taxes in the world, Alberta has no sales tax and the lowest corporate taxes in Canada). It is presented as a form of freedom, though it could be viewed as a form of bondage, a modern twist on that revolutionary cry from 250 years ago: "No Taxation Without Representation." Replaced by little taxation and little representation. Louisiana, another low-tax petrostate (10 percent of US oil production), has the second-highest poverty level in the country, and a long, very colourful history of political corruption.

Kevin Taft, former leader of Alberta's benighted Liberal party (last in power in 1921), argued that Alberta's oil industry effectively usurped the province's democracy. In two well-documented books—*Shredding the Public Interest* and *Oil's Deep State: How the petroleum industry undermines democracy and stops action on global warming—in Alberta, and in Ottawa*—he outlines the incremental capture of the government by oil interests.

One of Klein's campaign promises had been to get the province's fiscal house in order—to slay the deficit and pay off the province's considerable debt. He repeatedly announced that spending was "out of control." This was true, though not always for the reasons Klein articulated.

His administration focused on the health care system, which a government official said had seen rising costs of 12 to 13 percent for the previous decade and was unsustainable, though the figures for 1984–85 were 3.6 percent, and the two following years were 6.4 and 7.5 percent respectively. And this didn't account for inflation or for a growing population. But Klein was a gifted salesman, and the image of a bloated health bureaucracy became fixed in the public imagination. He closed three Calgary hospitals, eliminated thousands of nursing jobs, and froze doctors' salaries. There were cuts to education, social assistance, services for seniors, and the environment. By 1996, Alberta, often touted as the country's richest province, had the lowest health care costs per capita in Canada, and among the lowest in North America. In terms of public service, Albertans were the most poorly served in the country. Where had the money gone?

The answer was: corporate subsidies, which were the highest in the country. More money went to corporate subsidies than to social services, transportation, the environment, and culture combined. As Kevin Taft pointed out, what happened was "a massive transfer of public wealth into private hands, through subsidies."

The corporate subsidies were designed to attract and retain business. Alberta already had the lowest corporate taxes in the country, part of what was touted as "The Alberta Advantage." But more money went to subsidies than was received in corporate tax; the subsidies had become corporate welfare. Between 1986 and 1993,

Alberta collected $4.65 billion in corporate tax and gave $9.97 billion in corporate subsidies (excluding $5.6 billion in farm subsidies) , for a net drain of $5.3 billion. This benefaction continued under Klein under the banner of attracting industry to the province, which it did. Yet government's share of revenue declined. Oil prices had tripled, but the government's share dropped by 39 percent, and the Klein government received less than half the per-barrel revenue of the Lougheed government. There was little public outcry, in part because there was relatively little reporting on what was essentially a transfer of power.

In 2005, with the budget now balanced, Klein announced that every man, woman, and child in Alberta would receive a cheque for $400, "the result of burgeoning oil and gas wealth." Here was populism at its crudest and most feudal, the King, or in this case the King's minion, scattering his coin to the peasants' grateful applause.

* * *

A 2006 PROBE Research poll showed that 90 percent of Albertans either overestimated or didn't know how much the oil sands contributed to government revenue (in 2004–5, the oil sands represented 9 percent of oil and gas revenues, though they accounted for 25 percent of production). A majority (87 percent) felt the oil sands should do more to protect the environment. Klein had cut the environment budget and expanded the area the oil sands could develop. The mining operation already affected 3,000 square kilometres, and in situ operations secured leases to another 35,000 square kilometres.

In 2006, twenty-one years after he'd left office, Peter Lougheed visited the oil sands and weighed in on their development. "It's just a moonscape," he said. "It is wrong in my judgement, a major wrong, and I keep trying to see who the beneficiaries are ... It is not the people of this province, because they are not getting the royalty return they should be getting. So it is a major, major federal and provincial issue."

Under Klein, the provincial government was populated with former oil industry people, and government ministers left to go into the oil business in a process of cross-pollination. Patricia Black, energy minister from 1992 to early 1997, had worked in the oil business for fifteen years before entering politics. "My first love, right from the beginning when I came out of university," she said, "was the oil sands. To me, it was the ninth wonder of the world, and jewel of Alberta..." In 1996, three cabinet members and four Conservative members of the Legislative Assembly formed a numbered company to invest in oil and gas drilling, which could be viewed as a conflict of interest, given that policy could affect profit. Alberta environment minister Lorne Taylor opposed the 1997 Kyoto Protocol, spending $1.5 million on an ad campaign against it to counteract a May 2002 poll that stated 72 percent of Albertans supported the Kyoto Accord. (Prime Minister Stephen Harper would later call Kyoto a "socialist scheme," and his government would eventually withdraw from the protocol.) After the Alberta Department of Energy published a paper outlining how emissions could be reduced, the office that produced the paper was disbanded. Klein joked that global warming was caused by "dinosaur farts." With Klein, the oil industry was no longer accountable to government; its only accountability was to investors, most of whom were from outside the country (a report showed that 70 percent of oil sands production was owned by foreign investors).

One of Klein's campaign slogans had been "get government out of the business of business." But now business was in the business of government, a government made by and for the production of oil. As a result, between 1990 and 2014, in the wake of Rio and Kyoto, as the world was trying to wean itself off fossil fuels and reduce emissions, Alberta's emissions went from 175 million tonnes to 274 million tonnes. Out of twenty-eight Organization for Economic Co-operation and Development (OECD) countries, twenty had reduced per capita emissions between 2010 and 2019. Canada was among those that increased emissions, the third-worst offender. The Peaceable Kingdom had become, in the eyes of many environmentalists, a rogue state.

Through the Looking-Glass

WHEN OIL GOES from $100 a barrel to $40, when demand decreases, drilling activity diminishes in response. The market dictates activity. And we have seen this over the decades in the oil business. It is a boom-and-bust industry. The staunch belief that the market is the surest way to sort out the economy, that government interference only distorts market forces, was almost exclusively the domain of conservatives. Oil, with its lingering wildcat image, is often viewed as the ne plus ultra of free enterprise.

Though oil has rarely operated in a free enterprise model. In Russia, Iraq, Iran, Venezuela, China, and other oil-producing countries, it is mostly a state-owned and -run enterprise, either explicitly or implicitly. In Saudi Arabia, the United Arab Emirates, and Qatar, oil is the state. Globally, it is one of the most subsidized industries, receiving grants, loans, tax credits, and royalty reductions. The International Monetary Fund, which includes externalities—downstream costs to society—in its calculations, put the 2023 figure for global oil subsidies at US$7 trillion. The International Institute for Sustainable Development estimated that using 10 to 30 percent of fossil fuel subsidies would be enough to pay for a global transition to clean energy.

In the Glasgow Climate Pact in 2021, a commitment was made to phasing out "inefficient fuel subsidies": "Our analysis shows that many of these government measures were not well targeted, and while they may have partially protected customers from skyrocketing costs, they artificially maintained fossil fuels' competitiveness versus low-emission alternatives." Low-emission alternatives are often cheaper, with solar costing less than natural gas in most countries.

As an industry, oil is a curious mash-up of swaggering laissez-faire capitalism and (very) quiet socialism, with legitimate claim to both extremes. It is an industry that requires tremendous entrepreneurial energy but also needs subsidies to protect itself from the vagaries of an international market prone to manipulation.

Charles Koch, chairman and CEO of Koch Industries, primarily an oil conglomerate, has a net worth of $62 billion, and is a keen libertarian and a vocal critic of subsidies. In a 1978 essay in the *Libertarian Review*, he wrote, "The majority of businessmen today are not supporters of free enterprise capitalism. Instead they prefer 'political capitalism,' a system in which government guarantees business profits while business itself faces both less competition *and* more security for itself... There is some justification at least in the taunt that many of the pretending defenders of 'free enterprise' are in fact defenders of privileges and advocates of government activity in their favor, rather than opponents of all privilege."

This was particularly true of fossil fuel companies. Those indulging in "political capitalism" mushroomed in the decades after Koch wrote his essay (the Koch conglomerate eventually received more than $422 million in subsidies, according to the Checks and Balances Project). To receive these subsidies, the fossil fuel industry first lobbied governments, then infiltrated them, occasionally capturing them completely. "Political capitalism"—where government bestows favours on industry—evolved into "corporate socialism," where industry became, to a degree, the government.

* * *

IN 1902, JOURNALIST Ida Tarbell asked H.H. Rogers, a director of Standard Oil, how he was able to "manipulate legislation." He responded, "[Politicians] come in here and ask us to contribute to their campaign funds. And we do it—that is, as individuals ... We put our hands in our pockets and give them some good sums for campaign purposes and then when a bill comes up that is against our interests we go to the manager and say: 'There's such and such a bill up. We don't like it and we want you to take care of our interests.'"

This wasn't unique to oil, of course, but no industry has pervaded government the way oil has. Reagan was favourable to its cause, opening up the country for drilling, and his successor, George H.W. Bush, a former oilman and occasional Texan, solidified the relationship with Big Oil. He took office in 1989, the year the Global Climate Coalition was formed, whose members included Exxon, Shell, BP, Chevron, the American Petroleum Institute (API), and the National Coal Association, among others. Co-opting the language of climate by oil companies became an Orwellian staple. The Global Climate Council, and the International Petroleum Industry Environment Conservation Association, led by Exxon, both followed.

Language became a potent weapon, usurping environmental tropes, sowing doubt, undermining science, and actively seeking to undermine reality. "Victory will be achieved," an API plan stated, when "those promoting the Kyoto treaty on the basis of extant science appear to be out of touch with reality." They worked to reshape the debate so that it was a stark choice between Climate and the Economy, a battle where the economy was real, with real effects on families, while climate change was a distant rumour.

Oil had been seeping into US government for a century, but it was George W. Bush, a failed oilman, who elevated oil to foreign policy. When he and Vice President Dick Cheney took office, Gore Vidal referred to it as the "oil and gas Bush–Cheney junta." Cheney had been chairman and CEO of Halliburton, the second-largest oil service company in the world, and one of the largest fracking operators. In 2001, Cheney commissioned a report on energy security, which warned of a global energy crisis. At the heart of that crisis was Saddam Hussein's erratic oil policies, and his veiled threats to take Iraqi oil off the market to destabilize prices. "Iraq remains a destabilizing influence to ... the flow of oil to international markets from the Middle East," the report read. "Saddam Hussein has also threatened to use the oil weapon ... The United States should conduct an immediate policy review toward Iraq including military, energy, economic and political/diplomatic assessments."

The first three came to pass. The invasion of Iraq was ostensibly about searching for weapons of mass destruction, though in his book *Fuel on the Fire*, Greg Muttitt cited declassified files, concluding, "The most important strategic interest lay in expanding global energy supplies, through foreign investment, in some of the world's largest oil reserves—in particular Iraq. This meshed neatly with the secondary aim of securing contracts for their companies." Dick Cheney's former company Halliburton won US$17 billion in contracts for Iraq's oil reconstruction without having to bid on them. Cheney denied any ties with his former company, but an internal Pentagon email indicated that Cheney's office had been intimately involved in awarding the multibillion-dollar contracts. The email, dated March 5, 2003, two weeks before the invasion of Iraq, showed that Douglas Feith, Under Secretary of Defense for Policy, had approved a contract with Halliburton "contingent on informing WH [the White House] tomorrow." The email went on to say, "We anticipate no issues, since action has been coordinated with the VPS office." In all, $107 billion worth of oil contracts were awarded without open competition. The invasion of Iraq was lightly disguised as a military threat, but it was another Oil War.

Besides Bush and Cheney, the Bush administration's connections to oil included at least thirty people, including Condoleezza Rice, who was national security adviser then secretary of state, and served on the board of Chevron for almost a decade. Bush's secretary of commerce was Donald Evans, who had been CEO of the drilling company Tom Brown. Bush appointed Philip Cooney, who had been with the American Petroleum Institute for fifteen years, as chief of staff of the White House Council on Environmental Quality.

The Bush administration actively suppressed and sometimes altered scientific evidence that ran contrary to the industry's narrative. In 2003, *The New York Times* reported that White House officials tried to force the Environmental Protection Agency to remove information that confirmed human activity contributed to climate change, with the aim of stressing "a recent, limited analysis [that] supports the administration's favored message." Some of the changes were substantive, others were subtler and had to do with language, such as inserting the words *may* or *potentially* to create a sense of scientific uncertainty where there was clear consensus. In the end, the EPA deleted the entire section on climate change rather than deliberately misrepresent the science. Russell Train, a former EPA administrator under Presidents Nixon and Ford, wrote a letter to *The New York Times* saying that Bush's political manipulation of EPA findings was unprecedented, noting that the "interest of the American people lies in having full disclosure of the facts."

This was an era when facts were still something to be suppressed or massaged or undermined by doubt-inducing adverbs like *potentially* or auxiliary verbs like *may*. There would come a time when facts no longer had to be massaged or suppressed. Instead, they could be invented and mass-produced and disseminated at the annual Climate Change Conference: Renewable energy is an instrument of social control. Climate science is a communist hoax. Severe flooding in southern Brazil was God's judgment on a Madonna concert that contained "satanic content." All of this found a receptive audience as a disturbing portion of the country stepped through the Looking-Glass.

* * *

THE INFLUENCE OF oil on American politics reached a high point when President Donald Trump appointed former ExxonMobil CEO Rex Tillerson as secretary of state during his first term in office. In his book *Private Empire: ExxonMobil and American Power,* journalist Steve Coll described Exxon's relationship with American policy: "The corporation's lobbyists bent and shaped American foreign policy, as well as economic, climate, chemical and environmental regulation." This all happened behind the scenes, with Exxon keeping as low a profile as possible. "Exxon's foreign policy sometimes had more impact on the countries where it operated than did the State Department," Coll wrote. And now Tillerson was the State Department; oil could officially become foreign policy.

As CEO of Exxon, Tillerson had ties to both Vladimir Putin and Igor Sechin, head of Rosneft, the Russian oil company. Exxon and Rosneft had a joint venture in the Russian Arctic. America still had sanctions against Russia for its annexation of Crimea, and removing those sanctions would have benefited Exxon, yet Tillerson was in favour of keeping the sanctions in place (though he may have simply read the political mood—the vote to keep the sanctions was 419–3 in the House and 98–2 in the Senate). It was Trump who wanted to remove them, either dazzled by Putin's aura of gangster power or subdued by lurid *kompromat* or held hostage to business ties with Russia. How Tillerson's tenure would have played out for the oil industry turned out to be moot; he was fired after fourteen months in office, soon after calling Trump "a fucking moron" in private conversation.

Like Reagan, Trump appointed an anti-environmentalist to administer the Environmental Protection Agency. As attorney general of Oklahoma, Scott Pruitt had filed fourteen legal challenges against the EPA. As head, he cut the budget, approved twenty-four regulatory rollbacks, suspended the Clean Water Rule, and withdrew from the Paris climate agreement. Like Anne Gorsuch under Reagan, he resigned amid scandal, replaced, in a frying pan/fire scenario, by former coal lobbyist Andrew Wheeler.

Under Trump, the "capture" of government by private interests was complete. He leased millions of acres of public land for drilling, including a wildlife refuge, at below market prices, and declared that global warming was a hoax created by the Chinese in order to undermine American manufacturing. By noon on the day of his inauguration, the phrase *climate change* had been scrubbed from the White House website. Trump attacked more than a hundred environmental protections, rolling back the gains made by Obama, his political nemesis. He weakened rules that limited the flaring of methane on public lands, championed coal, and rebranded fossil fuels as "freedom gas." Trump was a dismal failure as both developer and president, and arguably as a human being, but he is dangerously effective at branding—though "freedom gas," like his casinos, failed to catch on.

The idea of freedom has long been a fundamental part of oil's appeal. It gave us fast cars and the open road, part of America's cultural ethos. Though the road is no longer open, our cities clogged, our commutes long and hostile. Like Reagan, Trump invoked a nostalgic America where everything was better—its roads open, its people freer, its prices lower, the middle class ascendant. Reagan's version was essentially Rockwellian, a pre-lapsarian America that was white and God-fearing and had only really existed on celluloid. Trump invoked a similar world (albeit even whiter) but replaced Reagan's folksiness with the irrational rage of an Old Testament God, more intent on destruction than resurrection. Reagan's sunny campaign boasted that it was "morning in America." With Trump, it is always twilight.

When Trump was asked if he would be a dictator during a second term, he replied, "Only for one day." And on that day, he would close the border and "drill, baby, drill." His two bedrock priorities, surpassing even his limitless appetite for personal vengeance, were keeping immigrants out and producing more oil and gas.

During Trump's first presidency, while he was invoking America's nostalgic, oil-fuelled past, China aggressively pursued the future, pouring government money into research and development of electric vehicles, solar panels, and battery manufacturing. China knew it couldn't compete with the West on internal combustion vehicles, and Japan dominated the hybrid market. But the manufacture of EVs was still in its infancy. By the fourth quarter of 2023, Chinese EV company BYD had become the biggest EV brand globally, eclipsing Tesla. The benefits to China were several: It was a boost to its dormant domestic auto economy, it helped mitigate China's alarming air pollution, and it decreased China's dependence on foreign oil. In 2023, China hit peak demand, with domestic oil consumption down 1.4 percent in 2024.

The New York Times reported that by the end of 2024,

China was producing 80 percent of the world's solar panels, controlled 60 percent of wind power manufacturing, and was making 75 percent of the world's lithium batteries. "China achieved this commanding position by taking nearly the opposite approach that Mr. Trump did ... The combination of Mr. Trump's apathy and China's investment means America is now rushing to catch up to China's lead in solar, wind, battery and electric vehicle industries. It is an open question whether it can get there." Trump set America back during his aimless gangster reign, but he managed to Make China Great Again.

In April 2024, Trump met with more than twenty top oil executives at Mar-a-Lago and made them an offer: If they gave him $1 billion for his third presidential campaign, he would roll back more than a hundred environmental regulations put in place by the Biden administration. This would be of net benefit to the oil industry and would happen on his first day in office. He ranted about wind turbines, which he claimed caused cancer, and offshore wind farms, which were "driving whales crazy." He also told the assembled that Ultimate Fighting Championship fighters should be in two separate categories: one for illegal immigrants, the other for Americans.

Karoline Leavitt, a spokesperson for the Trump campaign, defended the offer to oilmen in a statement, saying President Biden was controlled by "environmental extremists who are trying to implement the most radical energy agenda in history and force Americans to purchase electric vehicles they can't afford."

Trump promised to declare an "energy emergency" on his first day in office, in an attempt to bypass Congress and the nation's regulatory system. He was right that there was an emergency, though wrong to identify it as a lack of support for fossil fuels. The coal industry backed Trump's candidacy, hoping for more subsidies for "beautiful, clean coal."

On September 30, 2024, Britain shuttered its last coal plant, 142 years after its use of coal began, powering the Industrial Revolution. The phase-out largely happened under the Conservative government, with Labour supporting the move. But energy policy in the US is almost wholly political, a whiplash between two opposing philosophies, bringing green energy initiatives with the Democrats (President Biden's Infrastructure Bill included $73 billion for clean energy) and environmental rollbacks and oil-friendly policies with the Republicans.

In his second term, Trump said he wanted to foster "energy dominance" globally through increased production and export of fossil fuels. Trump has always been big on dominance, a concept that has infected his life, but the future is in renewables, an area where China dominates, in part because of Trump's willful ignorance.

* * *

IN CANADA, THE relationship between oil and the federal government has gone from the visceral, lasting hostility that Pierre Trudeau fostered, to the warm embrace of Stephen Harper's administration, to Justin Trudeau's extended fence-sitting. At the federal level, there wasn't the degree of regulatory capture that Alberta experienced, though oil made significant inroads, affecting policy and occasionally dictating it.

The Energy Policy Institute of Canada (EPIC) was formed in 2009, and its goal, according to Thomas d'Aquino, its first chair, was "to develop an Energy Strategy for Canada," one that was limited to fossil fuels. In a later interview with RCMP investigators, d'Aquino cited several key strategies, among them the creation of "public sensitivity" through newspapers and magazines, and "build[ing] support by engaging the academic community, the ones writing about energy." This followed the American lead: The Koch Foundation boasted a "robust, freedom-advancing network" of almost five thousand academics in four hundred universities.

The academic community already had an organization engaged on the energy front. The Canada School of Energy and Environment (CSEE) was formed in 2007 with a $15 million grant from Stephen Harper's government, partly with the aim of rebranding the oil sands as "sustainable." Straddling EPIC and CSEE was Bruce Carson, a former senior adviser to Stephen Harper. He chaired the CSEE and was on the payroll of EPIC. Carson would eventually be found guilty on three counts of illegal lobbying and fined $50,000.

Carson had an interesting pedigree, in and out of court for bouncing cheques, defaulting on a mortgage, and failing to make lease payments on a car. As Kevin Taft outlined in his book *Oil's Deep State,* "In 1983, he was convicted on two counts of theft, sentenced to eighteen months in jail and lost his license to practice law. In 1990, he pleaded guilty to defrauding two different banks and a car rental company and, as a condition of his sentence, was required to undergo psychiatric treatment." Then he became a senior adviser to Prime Minister Harper.

While serving as executive director of CSEE, Carson returned to politics on occasion, helping with a federal election campaign and briefly working in the Prime Minister's Office. The line between politics and think tank had blurred. Carson attended the Copenhagen Climate Change Conference in 2009 as "Senior Advisor to the Deputy Minister of the Environment." The following year, CSEE hosted a three-day meeting in Banff with Jim Prentice, then federal environment minister, and other government officials. It was hosted by Carson's then-girlfriend, Barbara Lynn Khan, a former sex worker from North Carolina who had been convicted of money laundering and running a bawdy house ("the Sugar Shack") and been deported to Canada. Not long after, the sixty-four-year-old Carson broke up with Khan and took up with a twenty-two-year-old former escort.

Both CSEE and EPIC were formed as industry advocacy groups that co-opted the language and provenance of environmental research and oversight, touring the world as a warped ambassador, pushing oil interests, sometimes under the guise of green initiatives. While Carson ran the CSEE, he also accepted contracts to work for the Canadian Association of Petroleum Producers. Between July 2008 and November 2012, EPIC, as well as oil industry executives and other pro-oil groups, had 2,733 meetings with federal officials, including Members of Parliament. The 2012 Canadian Environmental Assessment Act adopted the industry's recommendations, sometimes verbatim, industry now writing policy.

This was clear with Bill C-38, which repealed Kyoto, weakened the National Energy Board by allowing cabinet to override its decisions, decreased public participation in environmental issues, and reduced the number of projects that would be reviewed. The Harper government also suppressed the findings of federal scientists, changed the regulatory process, stopped funding the Canadian Foundation for Climate and Atmospheric Sciences, and killed Environment Canada's Adaptation to Climate Change Research Group, the National Roundtable on the Environment and the Economy, and the Canadian Environmental Network.

After Carson was convicted of illegal lobbying, Madam Justice Catherine Kehoe wrote in her sentencing report: "It is especially egregious in the case of EPIC where Mr. Carson was representing a non-profit corporation set up to represent numerous major private Oil and Gas Energy Companies whose sole purpose was to develop energy policy for Canada for the commercial benefit of the companies, while the public, including other interested companies, environmentalists, etc. had no knowledge of what was transpiring behind the scene with Ministers, Deputy Ministers, and other senior officials in government, both federal and provincial."

The oil industry had already captured Alberta's government, and then expanded east, seeking to capture Ottawa as well.

* * *

IN 1920, BRITISH economist Arthur Cecil Pigou published his influential book *The Economics of Welfare*. In it, he developed the concept of externalities—the negative downstream costs of a market activity. In Pigou's view, subsidies would be given to activities that benefit the public good, and taxes levied on those activities with a negative social value. For example, a carbon tax on oil producers could offset the downstream costs of pollution, which affect society, and the extra expense of the taxes could spur innovation to decrease pollution. There are Pigouvian conservatives who espouse this view. In a 2021 op-ed in Salt Lake City's *Deseret News,* nineteen Republican members of the Utah legislature declared support for a carbon fee. "We believe the best way to cut emissions is with a price signal to the private sector, which lets competition and innovation find the solutions." For Pigouvian conservatives, this is the freest of enterprises, one where the market operates at its most efficient, driven by competition and innovation. But it is mostly environmentalists who have emerged as Pigouvian conservatives, while Big Oil favours "political capitalism."

This is the topsy-turvy world of modern energy. The oil sands are "clean energy." There is "clean coal." Environmentalists are communists. Subsidies are free enterprise. Universities lobby for oil corporations. Those who believe in climate change are fantasists, or the Chinese invented it, or only God can change the climate, as Oklahoma senator Jim Inhofe, chair of the Senate Environment and Public Works Committee, stated: "God's still up there. The arrogance of people to think that we, human beings, would be able to change what He is doing in the climate is to me outrageous."

The Looking-Glass world of energy was personified by COP28, the annual United Nations climate conference, which was held in December 2023 in the United Arab Emirates, the world's seventh-largest oil producer. The president of the conference was Sultan Al Jaber, CEO of

ADNOC, the state-owned oil company that had plans to dramatically expand production. Al Jaber was accused by the BBC and others of using the forum to promote oil and gas deals. Two weeks before the conference began, Al Jaber was quoted as saying, "There is no science out there...that says the phaseout of fossil fuel is what's going to achieve 1.5 [the maximum rise of 1.5°C, the forlorn hope of every annual conference]." The UAE has the sixth-highest per capita greenhouse gas emissions in the world, and between 2015 and 2022 its emissions increased by 7.5 percent, compared with the 1.5 percent global average. There were 2,456 oil and gas lobbyists in attendance at the conference, more than most countries' delegations, and it gave the world two weeks of aspirational goals, non-binding agreements, happy pledges, and misinformation (the earth is cooling; wildfires are the result of arson; oil and gas companies are leading the net-zero revolution).

It took thirty years of Climate Change Conferences for the words *fossil fuels* to appear in any resolution, a welcome addition, though not one that was legally binding. Oil had captured governments around the world, and has now arguably captured the UN conference on climate change as well, a bloodless coup. COP29 took place in Azerbaijan, the birthplace of oil, a petrostate that relies on oil for 90 percent of its exports. Wealthy nations pledged $300 billion annually to developing nations affected by climate change, though independent assessments put the required figure at $1.3 trillion per year to deal with the damage. Both figures are dwarfed by the IMF estimate of $7 trillion in annual subsidies for fossil fuel industries globally. Chandni Raina, India's representative, said of the $300 billion offer, "It's a paltry sum. I am sorry to say that we cannot accept it." With Trump's ascendancy to the throne, any American pledges were rendered moot, and the original initiative, like so many before it, had more to do with posturing than progress. Perhaps future COP gatherings can be held in the boardroom of ExxonMobil.

As Alice said, confronting the Looking-Glass, "If I had a world of my own...Nothing would be what it is, because everything would be what it isn't. And contrary wise, what is, it wouldn't be. And what it wouldn't be, it would. You see?"

Shifting Sands

IN 2005, I flew to Fort McMurray, to write about the oil sands for a magazine. The oil sands had exceeded Alberta's conventional oil production four years earlier, expanding faster than projected. The bitumen deposits lie beneath 142,200 square kilometres of northern Alberta, an area the size of New York State, though only 4,800 square kilometres of it is close enough to the surface to be minable. Deeper deposits are accessed by drilling, usually with steam-assisted gravity drainage (SAG-D) technology. Bitumen is the residue of marine life from more than a hundred million years ago. Those decomposed organisms formed light oil, but when the Rocky Mountains formed, the light oil deposits were pushed eastward into Alberta, along with oxygenated water. Trillions of bacteria in the water fed off the lighter hydrocarbon molecules, leaving, finally, a heavier, complex hydrocarbon that contained sulphur, heavy metals, resins, and asphaltenes. This bitumen was locked in sand and sandstone formations, where it became the sandy, tar-like substance that gave the oil sands its first name.

Traces of bitumen were found on tools used by Middle

Paleolithic Neanderthals forty thousand years ago. Ancient Egyptians used it in the mummification process. Bitumen was found near Babylon in 3000 BC, and was used as mortar in the walls of Jericho. In the first century AD, Roman naturalist Pliny noted bitumen's healing qualities, used to treat wounds, cataracts, and rheumatism, among other ailments. Indigenous people in northern Alberta used it to seal their canoes.

In 1921, the provincial government formed the Alberta Research Council with an emphasis on developing the oil sands as an energy source, but it wasn't until the 1960s that it gained momentum. The oil sands were a remarkable feat of engineering, one that required faith, ingenuity, perseverance, and massive capital deployed in a remote, harsh climate. For the mining operations, first the forest has to be cleared, which is done with D11 Caterpillars that weigh a hundred tons and clear the trees with a seven-metre blade. The bituminous sand is then taken to a separator, where two tons results in one barrel of oil. The process requires water, chemicals, and huge amounts of natural gas; in 2023, the oil sands used 6.8 billion cubic feet of natural gas per day (more than four times the Canadian residential sector's use of 1.5 billion cubic feet), which accounts for more than a quarter of the country's total gas production. Bitumen is among the least efficient fuels, as measured by energy return on investment (EROI), the energy spent extracting, processing, and distributing any energy. Conventional oil currently has an EROI of up to 30. (By contrast, hydroelectric and nuclear power have the highest EROI, ranging from 30 to 100. Wind power is between 18 and 50.) But the oil sands have an EROI of between 3 and 5. Only fracking, with an EROI

ON OIL 71

of 2 to 3, is lower. Over the decades, with fossil fuels, we are spending more energy to get less energy.

By the end of the seventies, many of the mammoth logistical issues had been dealt with and production was under way, modest at first, and very expensive, though always with the idea of expansion. But with the collapse of oil prices in 1982, the oil sands, already economically challenging, became folly once more. Money and workers drifted south.

They were revived by an unlikely ally. Prime Minister Jean Chrétien, who was Pierre Trudeau's lieutenant during the hated National Energy Program sixteen years earlier, came to Fort McMurray in 1996 to publicize extensive tax breaks for oil sands development: Companies could write off 100 percent of their capital costs in the year they were incurred. Chrétien was hoping to shore up his marginal Liberal presence in Alberta; in 1993, Anne McLellan won her Edmonton riding by one vote (amended to eleven in a recount), earning her the nickname "Landslide Annie" and giving the federal Liberals their first Alberta seat since 1968. She was given the cabinet post of minister of natural resources. The provincial government joined in by reducing the royalty rate to 1 percent (from 25 percent) until capital costs were paid off. This provided a further incentive to continue expanding. Tim Howard of the Sierra Legal Defence Fund said, "You have got to believe in the tooth fairy to believe that these companies are going to present their accounts in a way that shows a profit."

The results were immediate. Foreign investment flowed in, Fort McMurray boomed. Between 1995 and 2004, oil sands production increased by 133 percent, while provincial revenues from oil sands royalties shrunk by 30 percent. Revenue per barrel of oil equivalent declined by more than two-thirds, from $1.60 in 1995 to 50 cents in 2004. The federal government provided millions in research and development money for oil sands projects, offered tax breaks and royalty relief. The oil sands had become one of the most heavily subsidized fossil fuel enterprises in the world.

* * *

WHEN I ARRIVED in 2005, Fort McMurray was still booming, and the oil sands had entered a new phase. The idea of peak oil was in the air. Conventional oil in the US had already peaked, according to Matthew Simmons and other peak theorists. That year, Simmons published *Twilight in the Desert,* which suggested that the Ghawar field in Saudi Arabia—the world's largest conventional field— might have peaked as well, and that OPEC producers were overstating their reserves. The amount they can export is determined by the extent of their reserves, so there was incentive to exaggerate (in 1985, Kuwait reported a 50 percent increase of its reserves; six OPEC nations immediately followed suit, increasing their own reserves by between 42 and 197 percent). There had never been an independent audit and Middle Eastern reserves were notoriously suspect. Global reserves were in decline, and the American fracking revolution had yet to begin. Meanwhile, demand was increasing dramatically. China was rapidly urbanizing, and had gone from an oil exporter to an importer and was circling the globe making deals to supply its expanding middle class. India, Brazil, and the us all had growing consumption rates. There were predictions that only ten years of conventional oil were left—as well as estimates of more than a hundred years, a discrepancy that isn't unusual when it comes to fossil fuel statistics.

Oil was now being hailed as "energy security," a phrase born out of the 1970s embargo and sharpened with 9/11. It was an essential part of geopolitics, an instrument of foreign policy, and the oil sands represented the largest global reserves in a stable democracy. Venezuela had larger reserves but was crippled by inefficiency. Russia was inching toward being a rogue state and had a corrupt, Kafkaesque bureaucracy. Nigeria was beset by theft and civil unrest. The Middle East was unstable; when a car bomb went off outside the interior ministry in Riyadh, Saudi Arabia, in April 2004, killing four people and wounding 148, oil went up $1.87 per barrel within hours. That October, oil hit US$55.67 and analysts estimated that $15 of that was a "terrorist premium." But Alberta was arguably the most stable democracy in the world, ruled by two oil-friendly, right-wing parties for seventy years. So both Chinese and American interests began buying up oil sands leases, circling one another like fighters in the boreal forests of northern Alberta. Other countries followed—France, Britain, Norway, Japan.

* * *

THE LANDSCAPE THEY arrived to was otherworldly. North of Fort McMurray on Highway 63, known as "Suicide 63" for its many traffic deaths, the Suncor and Syncrude operations are essentially towns—over-lit, ceaseless, like Las

Vegas, containing thousands of hopefuls and billions of dollars. There were clouds of steam and smoke that had the viscosity of whipped cream, folding in opulent funnels trailing east. Flares were visible through the dense fog, dozens of stacks emitting steam, sulphur dioxide, nitrogen dioxide, and other elements that yielded, a former Fort McMurray mayor said, "the smell of money." Lights defined an exoskeleton that covered hundreds of acres, a scene that evoked both a space station and an industrial past. There were pyramids of sulphur, brilliant-yellow stockpiles. But even these scenes don't fully convey the scale of the operations. You have to look at aerial photographs that show the violent monumental gashes in the landscape, the tailing ponds that are among the largest local lakes, the industrial sprawl, all of it curving over the horizon. Almost 3,000 square kilometres of forest had been cleared, and 35,000 square kilometres of forest was licensed for drilling.

I was given the tour, one that hundreds of journalists, politicians, and investors from a dozen countries had taken, a tour that included a wildlife corridor that cuts through the tortured landscape and the recitation of environmental triumphs. The gigantism of the oil sands is a point of pride, presented in equivalencies of football fields or the distance to the moon or the GDP of developing countries. I spoke to Indigenous leaders, one of whom decried the environmental destruction and alarming cancer rates, another who had got with the program ("if you change the word *concern* to *opportunities*..."). I rode in a massive four-hundred-ton truck that carries bitumen to the separator, driven by an Ontario man who would eventually be out of a job as the trucks became automated.

There were eighty-one projects under way or in the planning stages when I arrived, and while Fort McMurray was booming, it had few earmarks of a boom town. In the 1970s, the boom had infected Calgary's civic personality and acted as an aphrodisiac, but the mood in Fort McMurray felt more subdued. The twenty-first-century boom looked less like Sodom and more like a government town.

The large oil sands companies had essentially become the government, fuelled by public money. Fort McMurray's Keyano College had a campus named for Suncor, providing training for oil sands employees. In 2004, Syncrude spent $107 million on local Indigenous industries, contributed to literacy programs, funded scholarships, community centres, archaeological studies, and environmental initiatives, and raised a herd of three hundred wood bison. The three largest companies at the time (Syncrude, Suncor, Shell) took on some of the personality traits of government. After incurring massive cost overruns that ran into the billions, their defence was vague. There was none of the bloodletting that free enterprise prides itself on, executive heads rolling to placate the market. Meanwhile, the Alberta government had increasingly retreated into a private enterprise model that touted self-sufficiency, cutting funding to hospitals, education, seniors, social assistance, and the environment. The oil sands companies were becoming bureaucracies—bloated, inefficient, supported by tax dollars, but providing jobs, education, and environmental management.

It is the last category that proved most problematic. There had been concerns voiced about the environmental impact, but it wasn't until November 2005 that the Pembina Institute published its extensive analysis "Oilsands

Fever: The Environmental Implications of Canada's Oilsands Rush," arguing that the pace of oil sands development was far exceeding the government's ability to manage the environmental impact. Environmental oversight was convoluted and tainted by the fact that almost all data came from the oil companies themselves. The environmental management was under the auspices of the Alberta Energy and Utilities Board, the provincial Ministry of the Environment, and Environment Canada. To proceed with any new oil sands project required a provincial assessment, but it could set the scope of the assessment, say, studying the impact on a single creek, as one assessment did.

Neither the provincial nor the federal government looked at the cumulative, long-range effects of oil sands development. That job fell to the Cumulative Environmental Management Association, which got 90 percent of its modest budget from industry. It worked out of a house in Fort McMurray with a small staff, delicately examining some of the oil sands' fundamental hurdles: that it is both energy and resource intensive, and has significant greenhouse gas emissions, 2.2 times the emissions of the average barrel of North American oil. At the time, Syncrude was recycling 75 to 80 percent of its water, but it still took 31 million cubic metres of water out of the Athabasca River in 2004 (down from 40 million in 2002). Separating the bitumen produces tailings, some of which settle out in the water. However, 10 to 40 percent don't settle, and these fluid tailings take centuries to finally subside. The ponds contain naphthenic acids, cyanide, phenols, arsenic, lead, zinc, and other toxins, and don't freeze in the winter, so wildlife are attracted to them. In 2008,

1,600 migrating ducks died after landing in a Syncrude pond. As with oil spills in previous decades, the image of dead and dying birds resonated with the public in a way that statistics hadn't.

There is also the problem of air pollution emanating from the ponds. They emit hydrogen sulfide and nitrous oxide as well as methane and carbon dioxide. The industry estimated that 10 percent of their total greenhouse gas emissions were from tailings ponds. In 1973, before Syncrude was in production, Alberta Environment published a report that flagged tailings ponds as the largest environmental concern for the new industry. But cleaning up was voluntary in 2005, and would be until 2009, when the Energy Resources Conservation Board (now the Alberta Energy Regulator) finally produced standards for tailings management. Oil sands companies were required to capture 50 percent of all fluid tailings by 2013. That year, Premier Alison Redford announced that tailings ponds would "disappear from Alberta's landscape in the very near future."

But the ERCB immediately began accepting tailings plans that didn't comply with their own directive. When the regulator very quietly released its report in 2013, virtually every company had failed to meet the directive, and there were no consequences for non-compliance, no fines or penalties were imposed. Instead, the ERCB said its 2009 targets had been "overly optimistic."

An industry paper indicated a plan to let the tailings ponds accumulate until 2037, when there would be approximately 1.5 trillion litres of toxic waste. By then, perhaps, the technology to solve the problem might exist. But innovation is fuelled by investment and the possibility of profit, and in 2037 investment may not be directed at what is perceived as a dying industry.

More than 300 square kilometres of northern Alberta consists of tailings ponds, and they are growing annually. Every barrel of oil produces 1.5 barrels of tailings, and the oil sands produce more than three million barrels per day, with projected production increases in the coming years. There are currently more than 1.18 trillion litres of tailings, and a report from the Commission for Environmental Cooperation warned that they were leaking into groundwater. Federal inspectors examined the sites and determined that "when deleterious substances were found in groundwater samples, enforcement officers could not determine if they came from natural or anthropogenic" sources.

There was relatively little public response, in part because it wasn't widely reported and in part because we have become inured to environmental outrages.

The Alberta auditor released a report that estimated the cost of reclamation for the oil sands sites at $20.8 billion. There was $1.57 billion held in securities for the job. What protects the Alberta taxpayer from being stuck with the cleanup bill is the hope of an efficient technology in the future and the value of the oil sands themselves. But we are entering the musical chairs era of oil; who will be safely sitting and who will be stranded when the music stops? In their favour, the oil sands are vast and located in a stable democracy. Working against them is the fact that oil companies often prefer stable dictatorships to stable democracies. It can be easier to make money in Equatorial Guinea, ruled by a corrupt kleptocrat for decades, than in Alaska, with its (still quite lax) regulatory hurdles and environmental laws.

There is also the high cost of oil sands extraction ($45.92 per barrel at last count, compared with $5 for Saudi Arabia)—though it has come down dramatically over the years (it was $77.52 in 2015)—and the relatively low price it fetches in the marketplace. The benchmark for bitumen is Western Canadian Select, which can be anywhere from $2 to $55 less than West Texas Intermediate, the US benchmark. The reasons for this have to do with lack of transportation infrastructure, limited markets (almost exclusively the US), sulphur content, and processing costs. Globally, bitumen is the unwanted, slightly embarrassing cousin of crude oil. Other oil-producing countries have repressive or unpalatable political regimes but have lower extraction costs, and their oil commands higher prices. So determining who survives may come down to a battle between economics and politics. In their previous matchups (US–Saudi Arabia, Europe–Russia, et al.) economics remains undefeated. Both the US and Canada protested Saudi human rights abuses and the state-sanctioned murder of Saudi journalist Jamal Khashoggi, but both countries continued to import oil from Saudi Arabia (it remains the third-largest supplier to both eastern Canada and the US). Europe wished it wasn't helping finance Russia's war with Ukraine but had few energy options. Oil makes for inconvenient bedfellows.

* * *

IN 2022, SEVENTEEN years after I'd visited Fort McMurray, I returned to Calgary to talk to researchers who were dealing with innovation and conservation in the oil sands. By this time, production was much higher, though tax revenue from the oil and gas industry had shrunk; it was now a smaller part of the economy but a larger part of our environmental liability. The oil sands had made gains in reducing and recycling water, had lowered its per barrel emissions, had made some progress on tailings, but because it continued to expand, its carbon footprint had increased, and it remained the fastest-growing source of emissions in the country.

The oil sands were no longer at the centre of geopolitics as they had been in 2005. The US had discovered fracking, and OPEC was still going strong. The world wasn't running out of oil. It still represented energy security, but mostly it had reverted to just being money. Most international companies had exited the oil sands (Royal Dutch Shell, Marathon, Norway's Statoil—since renamed Equinor, which sold at a loss, Imperial Oil, ExxonMobil, Total, Repsol, and Koch Industries), most selling because of high production costs and low oil prices, though the public smudge of dirty oil didn't help.

The result of the exodus was that ownership was now much more concentrated. Six companies (Canadian Natural Resources Ltd., Cenovus, ConocoPhillips Canada, Imperial, Suncor, and MEG Energy) controlled 95 percent of oil sands production. A 2021 report from the Alberta Energy Regulator noted that Canadian-headquartered companies owned 85 percent of oil sands production. A report from three environmental groups—Environmental Defence, Équiterre, and Stand.earth—reported that the oil sands were 70 percent owned by foreign companies and shareholders. Both of these statements were demonstrably true. While the companies were headquartered in

Canada, they were majority foreign owned. The environmental report, which took its data from Statistics Canada, annual reports, and Bloomberg, argued that the majority of oil sands profits left the country, a familiar model in the developing world.

* * *

SEVENTEEN YEARS EARLIER, the fear was that global oil supply was peaking. Now it was demand that was predicted to peak, as early as 2030, according to the International Energy Agency. The oil sands had become an environmental *bête noire*, not just in Canada but globally, a punching bag for celebrities and the subject of dire predictions: If all other industry in Canada went to net zero, the output from the oil sands alone would put us 32 percent above our 2050 emissions targets, and that assumes a further 30 percent reduction in emissions per barrel.

The oil sands were on the cusp of their most profitable year in history, with profits from the largest six companies calculated at $35 billion, but they were still heavily subsidized—though there is little agreement on what constitutes a subsidy. The industry declared that it had received no government subsidies. Cenovus CEO Jon McKenzie said, "We certainly hear political rhetoric with regard to oil and gas subsidies, we're really not sure what it means because, again, we're not really aware of any oil and gas subsidies for the industry."

At the other end of the spectrum, the International Monetary Fund, which factors in tax measures, reduced royalties, as well as externalities, calculated the 2020 subsidies to Canadian oil and gas to be $81 billion. Between the extremes offered by industry and the IMF, the OECD had the Canadian figure at $4.5 billion, while Environmental Defence calculated $18 billion. Oil Change International reported that among G20 nations, Canada had the highest subsidies for oil, gas, and coal ($14 billion per year between 2018 and 2020) and the lowest subsidies for renewables. Everyone connected to the oil and gas world has an increasingly pointed agenda, and statistics tend to vary widely, even aggressively. Oil is a belief system as much as a quantifiable corporate enterprise, and there are believers, blasphemers, and penitents, each with their own convictions and statistics.

* * *

THE UNIVERSITY OF Calgary has 280 researchers working on energy solutions, trying to reduce the province's carbon footprint. I spoke to one of them, Dr Steven Bryant, who was the Canada Excellence Research Chair Laureate in Materials Engineering for Unconventional Oil Reservoirs, an ungainly title that dealt with reducing the carbon footprint of the oil sands. He is a Tennessee native, tall and lean, wearing cowboy boots, looking more like a Nashville session musician than a scientist.

"Right now," he said, "Alberta is this crucible for all the problems facing society in the next century. And some of those problems are in conflict with one another." A microcosm of the world's dilemma—the need to reconcile existential environmental threats with the economic challenges of weaning ourselves off fossil fuels.

The mandate for his researchers was "to bridge the gap between the current technological status of the oil sands industry and where it needs to be for a sustainable, globally competitive future." His lab has developed incremental technologies including nanotechnology that lowers per barrel emissions, but they have also looked at more radical solutions, like leaving the bitumen in situ and extracting energy through molecular transfer.

But any radical solution runs into the fiscal realities of oil sands infrastructure. The industry has spent billions building that infrastructure. "One of the things about the oil sands is right now the zeitgeist has shifted," Bryant said, "and folks aren't looking at building new anything. It's a very tough environment to go in and say, 'We can do things differently'." In 2021, Pathways Alliance, which represents the six largest oil sands companies, responsible for 95 percent of the industry, pledged to get their operations to net zero by 2050. But more than a year later, no plans had been announced, and the industry appeared to be going in the opposite direction, shelving $2 billion in green initiatives.

Shell had declared itself to be a leader in energy transition, but at an investor call in June 2023, Shell CEO Wael Sawan announced they would slow investment in renewables and their low-carbon business in order to boost shareholder returns. Suncor's CEO Rich Kruger concurred. "We have a bit of a disproportionate emphasis on the longer-term energy transition," he said. Suncor shelved a $300 million wind power project and a $1.4 billion cogeneration plant that would replace coke-fired boilers with natural gas. Companies emphasized the need to get back to their core business, which was oil, and to their core constituents—their shareholders.

One of the reasons there is little appetite for innovation

is that carbon capture and storage (CCS) has become the environmental holy grail for the oil sands. It is a process where emissions are captured at source and sequestered below the surface. It reduces CO_2, nitrogen dioxide, and sulphur dioxide emissions, but it is also energy intensive (the processes of separation, transportation, and injection), expensive, and difficult to scale up to the extent needed. And if the captured CO_2 is used for enhanced oil recovery—getting more oil out of existing wells—then it could add to emissions rather than reduce them. It has great appeal for the industry as it allows them to continue to increase the production of fossil fuels and to rebrand as a "carbon management industry."

It is also almost entirely funded by taxpayers. A report from the Canadian Institute for Climate Choices, a federal agency under the auspices of Environment Canada, warned, "Public investment in assets at elevated risk of being stranded in low-carbon scenarios could generate less economic and job benefit than investment in areas that could capture a share of growing, transition-opportunity markets." Nevertheless, the federal government pledged $12.4 billion in tax credits for CCS, while Alberta will pay up to $5 billion. The original estimate for the project was $16.5 billion.

In 2020, CCS projects captured about 40 million tonnes of CO_2 globally, about a thousandth of what we produced that year (38 billion tonnes). In May 2024, Capital Power, an Edmonton electricity generator, announced it would scrap plans for a $2.4 billion carbon capture plant, noting that it was technically viable but economically unfeasible, this despite having 62 percent of costs paid by provincial and federal governments.

Carbon capture can play a part in the solution, but it won't be the solution. "If I've learned anything about the climate challenge, the CO_2 mitigation challenge, the net-zero challenge," Steven Bryant said, "it's that it's got to be a collection of myriad solutions." There is no magic bullet.

* * *

DR STEPHEN LARTER is associate vice-president of research and innovation for the University of Calgary. "If the oil companies are serious about decarbonizing," he told me, "we'll see that revenue stream going into mitigating existing emissions, new clean energy technology, and an urgent energy transition away from where they are today. If they're not serious about it, then we'll see stock buybacks and executive bonuses and all the usual stuff."

We got the usual stuff: $32 billion went to share buybacks (which help increase stock prices), $16.7 billion went to dividends for shareholders, executive compensation increased dramatically (Imperial Oil CEO Brad Corson's salary almost doubled, to $17.34 million), with only $500 million going toward reducing emissions in the form of carbon capture, or 1.4 percent of profits. All of this suggests that decarbonizing is a distant priority.

Larter said the climate challenge needed the level of focus and investment that was brought to the Manhattan Project during the Second World War. "The Manhattan Project went from zero to a bomb going off in four years, when little was known of nuclear processes at the project start. And how did they do it? It was government plus academia plus industry driven by a timeline." Current efforts look less like the Manhattan Project and more like Manhattan: noisy, unaffordable, riven by inequality and corruption, and always facing the threat of decline.

* * *

MUCH HAS BEEN made of Norway's prudent stewardship of its oil and gas resources (which are smaller than Alberta's), treating them as a finite, publicly shared resource and producing a sovereign fund worth $1.7 trillion. It is essentially a national investment account, with some of the money designed to mitigate the inevitable decline in oil and gas revenue, whether through reduced demand or exhausted fields. By contrast, Alberta was a multi-decade keg party, with years of roaring fun, crippling hangovers, and searing regret. It also has a fund, started by Peter Lougheed, and over the same timeline it has saved a more modest $21.6 billion. Enough for cab fare home, but not enough to clean up the mess.

Orphan in the Storm

A UNIVERSITY OF Alberta project determined that 619,503 oil and gas wells had been drilled in the province. Of these, at least 3,406 are orphan wells, meaning the owners are bankrupt or no longer exist. The Dickensian-sounding Orphan Well Association lists 14,856 sites (sites, wellbores, facilities, pipelines) that need decommissioning and/or reclamation. As well, there are 170,000 inactive wells (or 230,000, depending on the source) that could end up orphaned. At some point, all wells become inactive. They stop producing oil or gas, or the company no longer sees them as viable, or the company goes bankrupt.

There are orphan wells in British Columbia and Saskatchewan, and in southwestern Ontario. An explosion on August 26, 2021, in the town of Wheatley, Ontario, was believed to be the result of an abandoned gas well. It was the fourth gas leak in the area, and the explosion sent seven people to hospital and destroyed two buildings. Ontario has 24,000 abandoned gas wells. Records are imprecise, and the town of Wheatley was built on top of four abandoned wells, though the exact locations were elusive. With some wells, methane gas erodes casings and plugs over time, and in the case of Wheatley it dissolved the gypsum rock, which releases sulphate, which then reduces to hydrogen sulphide, essentially acting as a chemical reactor. Cleaning up the Ontario wells will cost tens of millions, and the province has little appetite or expertise for the job.

The Alberta Energy Regulator's estimate for cleaning up Alberta's wells was a minimum of $60 billion and as much as $130 billion, a figure they flirted with then backtracked on. As of July 2023, there was less than $295 million held in security for the necessary work. The cost of cleaning up a single well can range dramatically, from a few thousand dollars to millions. A well drilled in Peace River, Alberta, in 1917 cost $9 million to clean up a century later.

In October 2023, the University of Calgary's School of Public Policy issued a report titled "A Made-in-Alberta Failure" that criticized the current orphan well policy for its lack of transparency as well as undue industry influence on the regulator. The orphan well issue was "an immense environmental and financial crisis that has been unsuccessfully dealt with by various policies over several decades," the report read. Every Alberta government has pushed the issue down the road, and compliance has largely been voluntary. "In the absence of significant and immediate legal and policy reforms," the paper stated, "the coming years and decades will see the enormous environmental, social and economic costs of this regulatory failure fall on the province's taxpayers." It has already fallen on federal taxpayers; in 2020, Prime Minister Trudeau announced that the federal government would spend $1.7 billion to help clean up orphan wells.

One of the paper's authors, associate law professor Martin Olszynski, noted that every decision about orphan wells was made behind closed doors, despite the fact that taxpayers may be stuck with the bill, and those decisions have largely favoured industry, part of the "regulatory capture" that has plagued the province. "I think the regulator simply has essentially conflated: What is good for the industry is good for us," Olszynski said in an interview with *The Globe and Mail*, "and what is bad for the industry is bad for us." He and other academics called for a public review of the Alberta Energy Regulator.

* * *

BEFORE SHE BECAME premier of Alberta, Danielle Smith was employed as a lobbyist for the oil industry. She is a self-professed libertarian and has espoused freedom at every level—political, economic, personal (anti-vax, pro-choice, pro–gay marriage and –private health care), and has wandered into conspiracy territory (smoking cigarettes can reduce the risk of disease, Russia invaded Ukraine to fight neo-Nazis) and awkward lapses (roughly comparing the vaccinated to those who voted for Hitler). In her capacity as consultant, she wrote a letter to then–energy minister Sonya Savage advocating for a break in royalties for oil companies who cleaned up old wells. Under Alberta legislation, companies were already obliged to remediate any wells they drilled, though compliance was low. After becoming premier, she immediately implemented that policy, instructing her energy minister to "develop a pilot program to effectively incentivize reclamation of legacy oil and natural gas sites and enable future drilling." The last three words were perhaps the most critical. In her keynote address to the Canadian Executive Association, Smith said, "We don't need a just transition in Alberta because we don't intend to transition away from oil and gas."

The Liability Management Incentive Program would give the oil industry $100 million in tax breaks if they cleaned up wells that have been inactive for at least twenty years. This was more gift than grant, given that they were already legally obligated to clean up defunct wells. There were other gifts. A corporate "job creation" tax cut resulted in a $4.3 billion savings for oil sands companies; between 2019 and 2022, 3,452 jobs in that sector were lost, largely due to automation.

* * *

IN JUNE 2023, Danielle Smith announced a six-month moratorium on new wind and solar projects until a study could be done to assess their cumulative environmental impact—the effect on the province's energy grid, and what it would cost to deal with stranded solar and wind farms. Both solar and wind present their own inefficiencies and environmental impacts. But in the wake of giving oil companies $100 million to clean up orphan wells, her decision appeared to be both punitive and nakedly political. It isn't enough that you succeed, your enemies must fail.

The Pembina Institute estimated that the moratorium on renewable energy approvals affected 118 projects worth $33 billion, with planning, development, and construction creating the equivalent of twenty-four thousand jobs for a year. Smith defended the decision, saying it wasn't her idea, that she was acting on the advice of the Alberta Utilities Commission and the Alberta Electric System Operator. However, while they wanted clarity and legislative movement on how the renewable sector would affect agricultural land and the electrical grid, neither body had suggested a moratorium on approvals.

The result of the review was a 70,000-square-kilometre buffer zone that would prevent solar and wind projects from impeding a view of the mountains. Any development on agricultural land would need to demonstrate that it could coexist with livestock and crops, and the costs of reclamation would need to be provided ahead of development—two standards the oil industry is not held to. "Our government will not apologize for putting Albertans ahead of corporate interests." It has rarely put Albertans ahead of corporate interests, and in this case it is putting the corporate interests of oil ahead of the corporate interests of renewables, arguably to the detriment of Albertans.

The renewable energy sector in Alberta is the country's most robust and fastest growing. In 2020, it produced 210 megawatts (MW) of power from solar and wind. In 2023, the figure was 4,164 MW, compared with 649 MW for the rest of the country. Interestingly, Texas, the largest oil producer in the US, also produces the most renewable energy—more than California—despite determined efforts at the state level to thwart development. One of the key differences is that in Texas, landowners own their mineral rights, where in Alberta the province owns them.

Less oil money goes directly to Texas's state government than goes to the Alberta government, so there is less influence.

Alberta has the most sunshine hours of any province, and there is no shortage of wind, as anyone who has visited Lethbridge or experienced a chinook can attest. Smith had been openly critical of renewable power for both its unreliability and its impact on farmland, while advocating more gas-fired plants in a "natural gas province." But employment in oil and gas is declining; in 2019 there were 142,012 employed, falling to 128,292 four years later. Part of this was a result of the COVID-inspired drop in demand, but the industry is also increasingly automated, and conventional oil and gas production peaked in 1998.

Renewables made up 15 percent of Alberta's installed generation capacity in 2018; this figure had increased to 31 percent by 2023. In 2022 there were 183 renewable projects under way, valued at $116 billion, and three-quarters of the country's wind and solar were installed in Alberta. Europe is moving even faster. The year 2022 was the first in which the European Union used more renewables (22 percent) than gas (20 percent). Germany got 56 percent of its energy from renewables. In late 2024 there were two terrawatts (two trillion watts) of solar power installed globally, and more was installed in 2023 and 2024 than in all previous years combined.

Renewable energy has quickly become much cheaper and more efficient, while improvements in fossil fuel efficiency have been incremental. The fossil fuel industry is no longer looking to innovate. Instead, it has hunkered down, waiting for the inevitable storm.

ON OIL   93

* * *

EVERY CIRCUIT PREACHER knows that Jesus may get them into the tent but the devil is the main attraction. Nothing unites like a common foe. In 2019, then–Alberta premier Jason Kenney set up the Canadian Energy Centre (CEC), a war room with a $30 million annual budget, to wage war on opponents of oil and gas. An early target was the animated children's movie *Bigfoot Family,* which streamed on Netflix. On its website, the CEC wrote, "Brainwashing our kids with anti–oil and gas propaganda is just wrong—and Netflix needs to know that!"

The CEC went after environmentalists, which it alleged were largely foreign controlled. Kenney said environmentalists began to target the oil sands after a 2008 meeting of "special interests" held "at the Rockefeller brothers' office in Manhattan." "You can't make this up, it sounds like a John le Carré novel or something," Kenney told a party convention.

You can make it up, it turned out. The CEC hired accounting firm Deloitte to investigate foreign money behind environmental charities. It found that between 2003 and 2019, those charities raised $8.1 billion, and of that, only $37.5 to $58.9 million was "foreign funding directed to Alberta resource development opposition." A percentage that was far lower than foreign ownership of the oil sands.

The CEC was relatively dormant after its expensive, debacle-filled start, but was revived under Danielle Smith in 2023 with a $22 million ad campaign. One ad read, "When America chooses Canadian oil, it helps provide reliable, responsible, and affordable energy that is committed to environmental excellence." America has a fluctuating interest in "environmental excellence" on its own soil, and historically no interest whatsoever on foreign soil. We are already America's largest supplier of oil. What the ad campaign spoke to was the looming vulnerability of the oil sands.

The threat of stranded assets is oil's dark subtext. In a research paper titled "Reframing Incentives for Climate Policy Action," a group of academics, largely from Cambridge, predicted $11 trillion worth of stranded oil and gas assets by 2036 if net-zero targets are met. At greatest risk globally, they argued, were the Canadian oil sands, largely due to the cost of extraction and high emissions.

The Alberta Energy Regulator currently holds $1.71 billion in security to clean up the oil sands and coal mines, against an estimated $57.3 billion in liabilities. Oil sands companies haven't contributed to the cleanup fund in more than a decade. Ryan Fournier, press secretary for Alberta environment minister Rebecca Schulz, said, "Oil sands are still relatively young in their life cycle ... In the coming years, the oil sands will provide more financial security as required by the program." The oil sands are fifty-eight years old, if we count from their baptism by J. Howard Pew in 1967, and may enjoy a retirement like many of us—approaching seventy, our best days behind us, facing the threat of cognitive decline, thinking we should have saved a bit more. It speaks to the extent of regulatory capture that the environment ministry, ostensibly tasked with the welfare of its citizens, is acting as an apologist for the oil sands in a time of record profits.

The IMF released a paper warning of chaos for oil and

gas economies that fail to diversify. "Countries and businesses reliant on these [oil and gas] markets must formulate policies to address this transformation, including the development of renewable energy. To jettison their hidebound economies, which have led to low productivity and waste, oil-rich economies should commit to reforms that lessen obstacles to innovation and entrepreneurship." Yet, in these late innings, Alberta is slouching toward 1973, hardening its status as a petrostate. In Ralph Klein, the oil industry had a Useful Idiot; with Danielle Smith, they found a Happy Warrior.

Smith's stalwart defence of oil and gas has been presented as a defence of the Alberta "way of life," the two conflated over the decades. The entrepreneurial energy that founded that industry—both conventional oil and gas and the oil sands—was remarkable. And there were decades when oil and gas represented both progress and freedom. But now they represent decline and destruction, and that entrepreneurial spirit has shifted to renewables. Like Trump, Smith invokes another time: the romantic/heroic era of oil. It used to be that the future was every politician's favourite tense, but as the future looks increasingly bleak, the past has become the way forward for populist politicians.

Nationally, Conservative leader Pierre Poilievre is opposed to pricing any form of pollution from the oil and gas sector, would kill the federal assessment act, the carbon tax, and the Oil Tanker Moratorium Act, accelerate fossil fuel production, and support more pipelines. He proposed a summer break on gasoline taxes for consumers between Victoria Day and Labour Day, saving drivers an estimated 35.6 cents per litre. Poilievre has railed against lobby groups as an assault on democracy, though in a one-year period (June 2023 to June 2024) he and/or his staff had forty-six meetings with oil and gas lobbyists. He has pledged to ban oil from "dirty dictators." "Buying overseas oil from polluting dictatorships is terrible for our environment." Though Saudi oil, while freighted with moral deficiencies, is much cleaner than the oil sands. Poilievre proposed more east–west pipelines with the aim of fossil fuel self-sufficiency, an idea that would have made more sense three decades ago but would now run into environmental protest, lawsuits, refinery capabilities, and the economic maze of international trade. The oil industry has made lofty promises about lowering emissions but is essentially filibustering until a friendly government arrives. Poilievre has vowed to remove "red tape" and allow for more development. The oil sands have pushed their net-zero targets to 2050, a comforting abstraction for the industry, though a point where we may have passed a lethal 3°C rise in global temperature. Like the US, we will see a whipsawing of environmental regulations depending on who is in power. Poilievre had to cancel several "axe the carbon tax" rallies in BC due to the threat of wildfires. There was a time when this might have been viewed as ironic, but the Age of Irony, which relied on a shared reality to work, has long passed.

* * *

INTERNATIONAL CLIMATE GROUPS and Cambridge academics warn of stranded assets as the world approaches its net-zero targets, but in doing their own risk analysis, oil companies have very different math. For one, there isn't much compelling evidence the world will reach its net-zero targets by 2036 (2023 saw a global rise of greenhouse gas emissions of 1.1 percent). And an analysis of the investment and political climate might show both to be favourable for oil companies for some time. There is now an aggressively oil-friendly regime in Washington, and a better than 50 percent chance there will be one in Ottawa (at the time of writing, Pierre Poilievre enjoyed a twenty-point lead in the polls). There is a 100 percent chance of Alberta's continued largesse. Many oil-producing states are under authoritarian rule (Russia, Saudi Arabia, UAE, Venezuela, Brazil, Nigeria, Equatorial Guinea, et al.) and few face any real environmental constraints. Their biggest enemy is the market. They all have a strong incentive to resist the pivot to greener technologies.

And while there is a lot of talk about divestment from fossil fuels, there is less action. In 2020, Larry Fink, CEO of BlackRock, the world's largest private investment fund, with more than $10 trillion under management, announced that "climate risk is investment risk." It was widely seen as a call to divest from fossil fuels, though BlackRock retained significant investments in oil sands companies and privately voted against climate change resolutions at those companies. In 2021, Fink said, "Nothing is more greenwashing than divestiture. Because it doesn't change the [carbon] footprint of the world." His point was that divestiture didn't affect demand, only supply, which would exacerbate inflation. We need to decrease demand, but like BlackRock, we are conflicted: We want to reduce our carbon footprint but fly to Phoenix in January. We want to reduce emissions but can't afford a Tesla. In July 2023, BlackRock announced that it had appointed Amin Nasser, head of Aramco, the world's largest oil company, to its board.

So the reports pile up of a grim future with melting glaciers, rising seas, mass migration, rampant wildfires, and desertification, but there is still little meaningful pressure to change. From a financial perspective, there are no consequences, and often benefits, to maintaining the status quo. Royal Bank, Canada's largest, is proud of its ESG (Environmental, Sustainability and Governance) record. In its 2021 report, RBC noted it was ranked second in diversity and inclusion, had issued a green bond, "established the Climate Strategy Steering Committee and ESG Disclosure Council," and pledged "Net-zero emissions in our lending by 2050"—the last two, like so many green initiatives, purely aspirational. RBC is also the single largest lender to the fossil fuel industry in the world. This expansive grey area between intention and action has become yet another petroleum by-product.

* * *

GLOBALLY, THERE ARE an estimated 29 million abandoned wells. Orphan wells can leak methane and hydrogen sulphide into the atmosphere and into aquifers. In 2018, there were 3.2 million abandoned oil and gas wells in the US, leaking an estimated 281 kilotons of methane, according to the EPA, though it said the actual figure could be three times that. It estimated that Canada was leaking 10.1 kilotons of methane from its inactive wells. The global estimate is 2.5 million tonnes of methane escaping annually from inactive wells. A rise in oil and gas bankruptcies (they rose 50 percent in both Canada and the US in 2019) produces yet more abandoned wells. Long after it's no longer fuelling our lives, oil will remain a part of us. Evidence of its comforting, violent reign will be spread across the world for generations.

* * *

IN A SHORT *History of Progress*, Ronald Wright describes the tragedy of Easter Island. Situated in the South Seas west of Chile, Dutch sailors reached the island on Easter Day, 1722. What they found was surreal: a treeless island that had hundreds of massive stone monuments, some of them nine metres tall and weighing almost a hundred tons. The Dutch wondered how they could have erected these stone statues with no trees or rope to move them into place.

There had been trees on the island when the inhabitants first arrived, in the fifth century AD. By 1400, there were ten thousand people on the small island (165 square kilometres), divided into clans, each of which honoured its ancestors with stone statues. The statues became a competition, and the monuments got larger (the largest was twenty-two metres tall, weighed 200 tons, and, significantly, was never erected). Trees were cut down to transport the increasingly large stones. At some point, someone cut down the last tree. The island was small, so this wasn't something that happened offstage; the islanders could see their trees disappearing in real time. The rats they had originally imported for food ate the seedlings, so there was no new growth. And without the trees to root the soil, it was carried away by wind and washed away by floods, so agriculture was effectively stalled.

Fifty years after the Dutch first discovered this odd

society, Captain James Cook happened upon them. By this point there were more than a thousand statues and fewer than two thousand people. There were warring factions, and they had toppled and beheaded one another's monuments. Eventually, not a single statue was left standing upright. There was no wood for homes and some of the people lived in caves. The society had regressed. Archaeologists Paul Bahn and John Flenley wrote that they "carried out for us the experiment of permitting unrestricted population growth, profligate use of resources, destruction of the environment and boundless confidence in their religion to take care of the future. The result was an ecological disaster leading to a population crash...Do we have to repeat the experiment on [a] grand scale?...Is the human personality always the same as that of the person who felled the last tree?"

Easter Island wasn't the first society to destroy itself, and it certainly won't be the last. History is filled with examples of peoples who have mismanaged their environment and seen their society wither or be displaced. Six thousand years ago, the Sumerians lived in what is now southern Iraq. It was a very sophisticated culture, one that gave us time (hours, minutes, seconds), the first written works (the *Epic of Gilgamesh*), the first schools, and the concept of cities. The *Epic of Gilgamesh* contains stories that later appeared in the bible, the Garden of Eden and the Flood among them. The original survivor in the Sumerian version of the Flood is Utnapishtim, who is instructed to "tear down your house and build a boat, abandon possessions..." And, like Noah, he is told to "look for life . . . Take up into the boat the seed of all living creatures."

There is evidence that this literary flood may have been describing real events. The Sumerians lived on a flood plain, and the area upstream had been deforested, leaving them more vulnerable to flooding. They were a largely urban civilization, living in city states and farming the land around them. They invented irrigation, but this new technology came at a price. Eventually, over-farming began to take a toll. Wheat would no longer grow, so they switched to barley. Eventually, that failed as well as the land became degraded. "The Sumerians failed to reform their society to reduce its environmental impact," Wright states. "On the contrary, they tried to intensify production ... [T]he short-lived Empire of Ur exhibits the same behavior as we saw on Easter Island, sticking to entrenched beliefs and practices, robbing the future to pay the present, spending the last reserves of its capital on a reckless binge of excessive wealth and glory ... The result was a few generations of prosperity (for the rulers), followed by a collapse from which southern Mesopotamia has never recovered."

By 2000 BC, that part of Iraq had turned to desert, and it remains one 4,000 years later. The land never did recover.

* * *

BOTH CAPITALISM AND democracy deal in short-term time horizons. Public companies are enslaved by quarterly results, the need to impress investors on a metronomic basis or risk investment drying up. The average tenure of a Fortune 500 CEO is four years, putting constant pressure on successors to produce results quickly. Politicians have four-year windows to produce marketable, campaign-worthy results. As one politician noted, the problem with progressive environment policies is that the risks are short-term and the benefits are long-term, i.e., when someone else may be in power. For both groups, long-term planning can be largely performative.

Two groups that look at fifty-year time horizons are environmentalists and the Pentagon. Neither paints a rosy picture. A 2003 Pentagon report envisioned a world devastated by global warming, warning of shortages, chaos, and skies filled with greenhouse gases, some of these events occurring before 2030. "Humanity would revert to its norm of constant battles for diminishing resources," the report reads. "Once again, warfare would define human life." The last Oil War.

* * *

IT IS UNLIKELY that Alberta will be able to afford to clean up its stranded assets when the time comes (in 2023, the province spent more than $1 billion to clean up inactive wells, but it only reduced the total by 5 percent). The cost to the taxpayer will simply be too great. And with reduced oil and gas revenues, the government will have fewer funds at its disposal. A familiar scenario could unfold where foreign companies and investors quietly pack up their tents, leaving locals with the mess. There will be lawsuits, recriminations, promises of action. Those who can afford it will privately pay to have wells on their property remediated. Some of the land will likely be abandoned, leaving contaminated water and a desolate landscape.

The province's saving grace is its sheer scale and abundance of natural beauty. The oil sands are located in the largest boreal forest in the world, though relatively few visitors or residents have seen it. You can spend your life in Alberta and never see a pumpjack. Banff will remain a jewel (albeit one threatened by wildfires), the mountains will retain their majesty. But scale is less of a defence than it once was. The world is getting smaller. Quebec wildfires choked New Yorkers during an apocalyptic summer. Fires in British Columbia and northern Alberta affected Calgarians, Winnipeggers, and much of Saskatchewan. Rocky Mountain glaciers, which are melting at an almost geometric rate, have an impact on drought conditions hundreds of kilometres downstream.

In chaos theory, the butterfly effect has a butterfly flapping its wings and a tornado resulting weeks later, a hundred kilometres away. This was a poetic metaphor, though inching toward reality now. Globally, we all inhabit chaos theory, both environmentally and politically.

Oil Addiction

IN 2016, PRINCE Mohammed bin Salman announced, "We have developed a case of oil addiction in Saudi Arabia." This was an understatement; the modern state was created solely by oil, its economy financed by oil. Prince Mohammed's announcement came after oil prices dropped in 2014, resulting in a Saudi deficit of almost US$100 billion the following year. Oil accounted for 90 percent of the country's exports and 45 percent of its GDP. The prince's "Vision 2030" plan was an ambitious remaking of both Saudi society and its economy; women would enter the workforce, a tourism industry would be developed, the government would become more secular, billions would be invested in renewable energy, and a new $500 billion smart city—Neom—was planned.

The Saudi addiction to oil was glorious while it lasted. In 1973, the country had a population of seven million, a tiny kingdom half a world away from North America, yet it exerted enormous power. It was the leading proponent of the oil embargo that year, whose reverberations echoed around the world and are still being felt, evidence of oil's global sway. In the US, the lengthy lineups for gas were a staple of the evening news. The automobile still held an element of romance in the 1970s, still represented a particular kind of freedom that was now being curtailed, the world's most powerful nation held hostage by a tiny desert tribe.

In 2022, OPEC, led by Saudi Arabia, sharply cut oil production, driving up gasoline prices weeks before midterm elections in the US. President Joe Biden was outraged and threatened to ban weapon sales to Saudi Arabia, and discussed initiating a lawsuit against OPEC for collusion. In response, the Saudis threatened to dump US debt (it holds $119 billion), which would upset financial markets. Upon taking office in 2021 Biden had promised to make Saudi Arabia "a pariah" for its human rights record, which included the dismembering of Saudi journalist Jamal Khashoggi two years earlier. Months later, Biden was in Saudi Arabia fist-bumping Mohammed bin Salman, the friendship renewed. For fifty years, the US–Saudi relationship has been like that of an old married couple who fundamentally loathe one another but can't afford to divorce. Despite now being the world's largest exporter of oil, the US still imports almost 500,000 barrels a day from Saudi Arabia, part of the complex financial and refining network that dictates global oil shipments.

Saudi reserves have an estimated lifespan of sixty years, so there is a race in Saudi Arabia to diversify. Oil still accounts for 80 percent of Saudi exports and 40 percent of its GDP, making it the third most oil-reliant country in the world, behind Kuwait and Libya. But it is aggressively building solar and wind capacity, with the aim of getting 50 percent of its electricity from renewables by 2030. Fellow OPEC members Qatar, Bahrain, and the United Arab Emirates are also embarking on mammoth diversification schemes (UAE Vision 2021, Abu Dhabi 2030), expanding tourism, investing in international education and renewable resources, as well as poaching soccer stars and marquee golfers and securing a scandal-plagued World Cup in Dubai.

Saudi Arabia is the elite embodiment of what has been termed "the resource curse," defined as "a paradoxical situation in which a country underperforms economically, despite being home to valuable natural resources." It can be the result of putting too much labour and capital into one resource that is vulnerable to price fluctuations, leaving a country economically exposed. It can also be the result of authoritarian rulers keeping the new-found wealth for themselves while the country's citizens live in poverty.

Oil has been a boon to Saudi Arabia, but it has created the inequality that is the hallmark of the resource curse. The Saudi royal family is the richest on earth, with $1.4 trillion in assets, while 20 percent of the country lives in poverty. Now it is moving away from a feudal model where everything is funded by oil, introducing taxation while at the same time providing less social infrastructure.

With oil came power, and addictions can be difficult to kick. But despite their outsized power, the Saudis were entirely dependent on foreign revenue in a single market. With Vision 2030, they will have less power but more freedom. Other countries defined by the resource curse aren't as lucky.

* * *

THE POSTER CHILD for the resource curse is Equatorial Guinea, a tiny country of 1.7 million people spread over 28,000 square kilometres on the west coast of Africa, wedged between Cameroon and Gabon. It was a Spanish colony until 1968, when Spain, still under the control of dictator Francisco Franco, handed the country over to Francisco Macías Nguema, who declared himself "The Sole Miracle of Equatorial Guinea." One of his first miracles was to murder or exile a third of the population. As Peter Maass reports in his book *Crude World: The Violent Twilight of Oil*, some were crucified, others rounded up in a soccer stadium and shot while a military band played the Mary Hopkin song "Those Were the Days."

Macías's nephew, Teodoro Obiang Nguema Mbasogo, was head of security, and in 1979 he staged a coup, deposing his psychotic uncle. Macías's trial was held in a movie theatre in Malabo with the accused suspended in a cage hung from the ceiling. He was found guilty and executed by a firing squad, leaving Obiang as the new ruler.

In 1995, oil was discovered in the Gulf of Guinea, an estimated 35 billion barrels, and Equatorial Guinea quickly became the third-largest sub-Saharan oil country, after Nigeria and Angola. Yet the country saw few benefits. The amount that foreign oil companies pay countries depends in part on how sophisticated and/or corrupt the country's leadership is. Oil companies pay a fee for the right to explore for oil, then royalties are paid on any oil that is discovered and taken out of the ground. Equatorial Guinea received relatively little on both counts because, according to the International Monetary Fund, Obiang either didn't understand how much he was giving up or didn't care.

One of the reasons he wouldn't care is that the money wasn't being paid to his country but to him directly. So the millions on offer were sufficient. Oil royalties were deposited into accounts in Obiang's name in the Riggs Bank in Washington, DC, finally reaching a total of $700 million. A Riggs employee went to the Equatorial Guinea embassy on two occasions and picked up suitcases that contained $3 million in hundred-dollar bills to deposit. Eventually, this came to the attention of US Senate investigators, who released a report in 2004 titled "Money Laundering and Foreign Corruption: Enforcement and Effectiveness of the Patriot Act, Case Study Involving Riggs Bank." Essentially, oil companies and an American bank were helping Obiang steal his country's money. The Senate investigation stated that "oil companies operating in Equatorial Guinea may have contributed to corrupt practices in that country by making substantial payments to, or entering into business ventures with, individual E.G. officials, their family members, or entities they control, with minimal public disclosure of their actions."

The reason there was minimal public disclosure was that neither the US nor Equatorial Guinea had laws requiring it. American oil companies paid more than $4 million for tuition so students from Equatorial Guinea could study abroad, though it turned out that these students were "children or relatives of wealthy or powerful E.G. officials," according to the Senate report. The country's own education system was among the worst in the world; 42 percent of children didn't go to school, and only 25 percent got as far as middle school.

Despite oil companies spending billions to build offshore drilling platforms and a refinery for liquid natural gas, very little of that money entered the local economy. Construction materials were brought in, and almost all the workers—in the plant, on the oil rigs—were foreign, mostly from India and the Philippines. Chinese oil companies brought in Chinese labourers and managers. They lived in imported prefab trailers and ate imported food.

By 2017, Obiang was the longest-serving president in the world, and one of the richest men in Africa. He owned a twenty-seven-metre yacht and a Boeing 737 with gold-plated bathroom fixtures. A Human Rights Watch report titled "Manna from Heaven?: How Health and Education Pay the Price for Self-Dealing in Equatorial Guinea" noted that the country's health and education systems were chronically underfunded and Equatorial Guinea was in violation of its human rights obligations. "Ordinary people have paid the price for the ruling elite's corruption," one of the paper's authors said. And by this time, its oil reserves were dwindling. The flow of capital was diminishing and would soon end.

Between 2000 and 2013, Equatorial Guinea took in roughly US$45 billion in oil revenues. Yet it is, according to the Gini Coefficient, which measures income and wealth inequality, the single most unequal country on earth. In 2019, it had the highest GDP per capita in Africa (US$36,270) and a poverty rate of 77 percent. Half the population didn't have access to safe drinking water, vaccination rates had declined, and health care was primitive and sporadic. The president's eldest son, Teodorín, managed to make $300 million as minister of agriculture, more than the combined budgets for health and education. The US brought a corruption case against him, and Teodorín had to forfeit $30 million in US assets, with $25.6 million returned to the country in the form of COVID-19 vaccines that went out to 600,000 citizens. The French government convicted him of embezzlement, guilty of transferring €110 million from the public treasury into his account, most of which went toward a €175 million shopping spree on a mansion, cars, and designer goods. The French government confiscated €150 million worth of his assets. Switzerland's investigation resulted in the seizure of eleven luxury cars and a €100 million yacht. The cars and yacht were sold and proceeds went directly to a social program in Equatorial Guinea. What social programs exist in Equatorial Guinea have largely been mandated by foreign powers. As punishment for these transgressions, Teodorín was made vice-president of Equatorial Guinea.

A lot of the oil money was invested in large infrastructure projects. There are new roads, sports stadiums, and a palace that occupies twelve square blocks. There is a new airport, five-star hotels, championship golf courses, and a new city being carved out of the rainforest. Very few locals were involved at any level in these projects. Because of the abysmal education system, there isn't the expertise for technical and management jobs, but even the manual labour was imported. The government owns the employment agencies, and prospective workers need to belong to Obiang's political party, and in some cases tithe some of their earnings to the government, in order to get a job. The various money laundering investigations that the US, France, and Switzerland conducted revealed that senior government officials make millions on contracts given to companies they own. Obiang and his family own stakes in the country's largest construction firms. There is no bidding process for projects.

American oil and gas companies that bribe officials or contribute to a foreign country's corruption don't face any real consequences. In 2017, President Trump struck down an anti-corruption rule that would require US oil and gas companies to disclose how much they pay foreign governments. The rule was designed specifically to combat the resource curse, to prevent autocratic leaders from taking payments for themselves rather than the national treasury, though in the case of Equatorial Guinea there was little difference between the two. American oil companies argued that the rule would put them at a disadvantage; other foreign companies weren't compelled to disclose payments. This wasn't true; both Canadian and European firms have to disclose that information. In 2016, Exxon, then led by Rex Tillerson, who was now secretary of state, was being investigated by Nigeria's economic and financial crimes commission regarding oil rights secured by Exxon in 2009, despite a bid that was $2.25 billion lower than a Chinese competitor's. Nigeria routinely makes the list of most corrupt countries on earth, a country that knows corruption when it sees it.

* * *

EQUATORIAL GUINEA HAD little in the way of environmental regulations. As a result, spills on offshore rigs were common, but, because the country is off the international radar, they went largely unreported. In 2021, French researchers studied ten years of images from the Envisat satellite and concluded there had been 18,063 oil slicks in the Gulf of Guinea between 2002 and 2012. Some of these were natural seepages from oil near the surface, but the vast majority were the result of shipping or production spills. The volume of oil was greater than the Deepwater Horizon spill, which leaked 4.9 million barrels into the Gulf of Mexico in 2010. In 2008, oil spills covered almost a thousand square kilometres in the Gulf of Guinea.

In 2022, ExxonMobil, one of the major companies in Equatorial Guinea, announced it would leave the country when its licence expires in 2026. Its output peaked in 2014, when it was pumping 300,000 bpd. By 2022, it was down to 15,000 bpd. Total production for the country was down to 95,000 bpd, its reserves projected to run out in 2035; with the quiet exit of major oil companies, revenue will fall steeply long before that.

There will likely be little demand for the golf courses and five-star hotels. When the party ends, there will be less rainforest, more environmental degradation, and the mocking reminder of abandoned drilling platforms. The economy could be gutted. In 1968, the year of its independence, cacao exports accounted for 75 percent of GDP, but only 8 percent of the country's land is now used for agriculture. Fishing was decimated by serial oil spills. This is one of the hallmarks of the resource curse: The society concentrates on a single resource and other industries wither. When the resource is exhausted, there is little to replace it. In a decade, Obiang's infrastructure, the monuments he built to himself, may resemble those stone heads on Easter Island.

* * *

TO THE NORTH, Nigeria sits as the former poster child for the resource curse. Foreign oil companies are leaving

ON OIL   113

Nigeria as well. Exxon, Shell, Chevron, Norway's Equinor, and Italy's Eni have retreated, selling their assets to local companies, due in part to high levels of oil theft. According to the Nigerian National Petroleum Corporation, they were losing 437,000 bpd. This is almost half Nigeria's production, which is roughly a million bpd, down from 2.5 million in 2011. Pipelines are blown up, fires started, tanker ships filled with oil set on fire. The United Nations reported that cleaning up the Niger Delta alone would take thirty years and more than $12 billion, which might be an optimistic figure. Nigeria has US$87 billion in debt, and 40 percent of its 235 million citizens live below the national poverty line. Oil accounts for 90 percent of its exports.

At the close of COP28, Nigeria's environment minister, Ishaq Salako, said, "Asking Nigeria, or indeed, asking Africa, to phase out fossil fuels is like asking us to stop breathing without life support. It is not acceptable, and it is not possible."

It is instructive that this was coming from the environment minister. Certainly, any shift away from oil will present challenges, though there was nothing legally binding in the COP28 resolution to phase out fossil fuels. But Salako's declaration was disingenuous. Eighty-five percent of Nigeria's oil wealth went to 1 percent of the population. Before oil was discovered, Nigeria had a robust and diversified economy. Currently, 56 percent of Nigerians—133 million of its 235 million citizens—live on less than two dollars per day and are defined as "multi-dimensionally poor," meaning they lack access to clean water and sanitation, health care, clean energy, and housing. Oil curtailed Nigeria's nascent democracy, killed critical parts of its economy, and promoted inequality. Pivoting from oil may ultimately benefit its citizens.

* * *

AN OECD PAPER on the resource curse analyzed twenty-four oil exporting countries (including Canada) and found "a 10-percentage point increase in the oil export share is associated with a 7% lower GDP per capita in the long run." The long run is the issue here; will there be one, and how long will it be? In the end, we are all cursed, cursed by oil's ease, seduced by its possibilities. It gave us freedom, warmth, hope, and ruin.

The Fracking Revolution

ON MARCH 2, 2016, one day after he was indicted by a federal grand jury for violating anti-trust laws, Aubrey McClendon, co-founder of Chesapeake Energy, America's second-largest fracking company, drove his Chevrolet Tahoe into a concrete bridge support on Midwest Boulevard in Oklahoma City. He was going 145 kilometres per hour, not wearing a seat belt, and there was no evidence he tried to swerve or brake. His death was ruled an accident. In both his flamboyant life and violent, enigmatic death, McClendon personified the fracking revolution, its manic rise and fatal flaws.

In 2006, the US imported 60 percent of its oil, mostly from the Middle East. By 2015, it was exporting oil and was the world's largest producer of natural gas, all due to fracking. Hydraulic fracturing, or fracking, involves drilling a horizontal well then sending fracking fluid under high pressure to fracture the rock and release its oil and/or gas. A version of the technology had been around for decades, but it wasn't until George Mitchell, a Texas geologist and entrepreneur, perfected the modern version that it became widely adopted. Mitchell had been trying a version of fracking since 1981, and after sixteen years it finally became a commercially viable technology. But the American fracking revolution didn't really begin until 2008, buoyed by low interest rates that followed the financial crisis. With fracking, American companies had access to billions of barrels of oil and gas that had been inaccessible with conventional drilling, much of it located in shale beds.

Fracking came with two significant downsides. First, it was environmentally more intrusive than conventional drilling. Between 2005 and 2015, 137,000 fracking wells were drilled in the US, and the high-pressure fluid used included 2.25 billion kilograms of hydrochloric acid, half a billion kilograms of petroleum distillates, which contain carcinogens, and 200 million kilograms of methanol, which may cause birth defects. Fracking fluid is proprietary, and companies don't have to disclose what is in it, but analysis by the Yale School of Public Health identified 157 fracking chemicals that are toxic. The reason companies don't have to disclose which toxic chemicals they are using is due to a provision to the Energy Policy Act of 2005. Endorsed by former Halliburton CEO Dick Cheney, who was vice president, the provision exempted fracking from federal regulation under the Safe Drinking Water Act. Fossil fuels represented energy security, which superseded environmental concerns or basic safety. Fear has been a valuable ally for oil over the years: fear of running out, fear of freezing in the dark, of economic collapse.

Each fracked well uses between 7.5 and 148 million litres of water, often drilled in states that are already experiencing water shortages (Texas, Colorado, and Pennsylvania among them). In 2014, fracking produced 53 billion litres of waste water, released 2.4 billion kilograms of methane into the atmosphere, poisoned water wells, and exacerbated drought conditions. More than 260 private wells were contaminated in Pennsylvania alone. Fracking companies were compelled to remediate sites after drilling was complete, but they rarely did, because they couldn't afford to. Which brings us to the second problem: Fracking is wildly unprofitable.

In part, fracking was a victim of its own success. The fracking revolution unleashed a huge supply of natural gas onto the market, driving the price down from $12 per million British Thermal Units to $3 per million BTU. But even at $12, it was difficult to make a profit, due to the nature of fracked wells. Conventional wells often pump steadily for years, but the diminishing return on fracked wells is steep; in their second year of operation, the average oil and gas output from a fracked well drops off between 60 and 80 percent. Then you need to lease more land and drill more holes. So the capital costs are high.

Money was very cheap, and private equity was happy to invest in what appeared to be a roaring enterprise. Chesapeake Energy was at the forefront of the revolution, driven by the expansionist zeal of Aubrey McClendon. He was eventually fired as CEO of Chesapeake, viewed as too reckless, and the company declared bankruptcy in 2020. In her book *Saudi America: The Truth About Fracking and How It's Changing the World,* Bethany McLean described how an inherently unprofitable industry was able to flourish for so long. "It's in some ways a tragic story," McLean said in a podcast interview, "because Chesapeake and its co-founder, Aubrey McClendon, really did, arguably, change the world. He had this idea that he could grow this company into the largest natural gas producer...and that fracking for natural gas was going to unleash this profound supply that was going to change America forever... But the company never made money." Even when gas prices were high, Chesapeake failed to make a profit. As McLean pointed out, the business model simply didn't work. The industry was propped up by cheap money and opportunistic banks. McLean calculated that Wall Street firms made more than $1 billion in fees raising equity and debt just for Chesapeake alone.

Between 2010 and 2020, shale drilling lost $300 billion, with Chesapeake accounting for $30 billion of that. For five consecutive years, the sixty largest fracking companies lost an average of $9 billion each quarter. It was a genuine energy transition for the US, one that came at great environmental cost, and it was consistently unprofitable.

The pandemic hurt oil prices, and shale drilling understandably faltered. But it enjoyed a renaissance afterward, in part because it had ceased to plough all its cash flow into further production growth. Oil prices soared to $100 a barrel, and more money went to dividends and share buybacks, which brought investors back. It was, the *Financial Times* noted in 2022, a stunning business reversal. But the *Times* also predicted that the shale revolution was winding down. The Bakken field in North Dakota was largely depleted. Pennsylvania had fewer than half the active wells it had had five years earlier. The Permian Basin in Texas was still going strong, but growing at a much slower rate, and yield per well was down.

As *The New York Times* noted, "Fracking has been, for nearly all of its history, a money-losing boondoggle, profitable only recently, after being propped up by so much investment from Wall Street and private equity that it resembled less an efficient-markets no-brainer and more a speculative empire of bubbles like Uber and WeWork." It was bolstered by $100 oil, but prices have slipped to less than $70 a barrel, once more making the economics challenging.

One of the benefits of the shale revolution was that it gave the US greater geopolitical leverage. The US became a net exporter of fossil fuels, and President Biden was able to deliver gas to Europe after Russia curtailed supplies following its invasion of Ukraine. And it gave him greater leverage in the prickly relationship with Saudi Arabia. He had entered office with the promise to crack down on fracking, and instead found himself quietly asking companies to produce more oil and gas in a classic confrontation between politics and *realpolitik*.

What was gained geopolitically was lost environmentally. A Harvard study of the health effects on people living near fracking sites showed a 3.5 percent increase in premature death. Other studies showed the prevalence of asthma, cardiology issues, and low birth weights. A 2010 documentary, *Gasland*, showed residents near a Pennsylvania shale field lighting their drinking water on fire.

In 2024, fracking was again a political issue, though it largely came down to the swing state of Pennsylvania in what was viewed as a tight presidential race. Pennsylvania was still fracking, though not at the hectic pace of a decade earlier. But pundits predicted the election might come down to that one state, so Trump loudly decried Kamala

Harris's earlier pledge to ban fracking ("She's totally anti-fracking. She's been anti-fracking and anti-drilling and anti–oil and gas practically since the day she was born"), while Harris reiterated her new stance ("As president of the United States, I will not ban fracking").

Trump linked oil and gas to jobs, though jobs in that sector are declining, but he also linked fossil fuels to freedom, America's bedrock belief. Geopolitically, this holds some truth. But with the shale revolution on the wane, the power and freedom that comes with oil is tenuous. Energy independence through renewables would be lasting, and would ultimately deliver more freedom.

It is testament to oil's stubborn mythology that decades of subsidies and the alarming debt incurred during the fracking revolution failed to dent its capitalist imprimatur. Renewable energy, which historically has received fewer subsidies (and continues to do so in Canada: $14 billion for fossil fuels versus $1 billion for renewables), remains "expensive," "unreliable," and "a socialist plot." When solar firm Solyndra declared bankruptcy in 2011 after receiving a $535 million government loan, Republicans used it as a political club for years. But fracking produced a flurry of billion-dollar bankruptcies, most of which flew under the radar. The tribal aspects of energy have less to do with what it is than what it represents. When Trump first announced that he would make America great again, he invoked America's glorious past, one where the people and picket fences were white, where crime was low and women stayed at home, where Elvis was the biggest threat to society. Oil and gas were fundamental to that world, to progress and freedom, two areas where America excelled. It was a potent message.

Many on the left recall a different past, a country riven with racism, bigotry, and sexism, an empire in danger of collapsing under its own weight. These two images are hard to reconcile. President Biden's Inflation Reduction Act, which included billions for clean energy, disproportionately favoured red states (Texas, Alabama, West Virginia, Ohio, Oklahoma). The state Republicans who vigorously opposed the act, declaring that clean energy was an elitist, globalist, left-wing plot, are now benefiting from it financially. Yet Republicans remain suspicious, as if aligning with the left on energy is a slippery slope; next up will be transgender policy, prayer in schools, abortion, and godlessness. The image of clean energy being progressive and dangerous, even when it provides jobs and taxes for the state, is a stubborn one. And we live in an age when image is far more powerful than reality.

The Rapture

FORMER PRESIDENT GEORGE W. Bush once stated that oil was the basis for Western civilization, and perhaps he was right. Oil continues to flourish, and both democracy and the rule of law—the former bases for Western civilization—are in alarming decline. Western civilization is beginning to look like the Book of Revelation, with its violent, ongoing apocalypse.

Revelation describes seven years of tribulation, where wars, disease, and natural disasters blanket the world. "A third of the earth was burnt up, and a third of the trees were burnt up, and all green grass was burnt up." A third of the sea becomes blood, a third of the creatures in the sea die, smoke covers the earth, and out of the smoke come locusts. A great beast arises from the sea and deceives the world. Revelation 13:5 tells us: "The beast was given a mouth to utter proud words and blasphemies and to exercise authority for forty-two months." Or roughly one term in office. Ominously, the beast also had a head wound, "but its fatal wound was healed. Rapt with amazement, the whole world followed the beast." Right into a second term.

Who can blame evangelicals for seeing Revelation mirrored in our times? The plagues and wanton excess, the inequality, the smoke and dying seas. Globally, 2023 was the hottest summer in history, unsurprisingly eclipsed by 2024, which was hotter still. Wildfires reached record levels (in 2021, 9.3 million hectares were lost to wildfires globally; in 2023, Canada alone lost almost twice that—18.5 million hectares). Swiss glaciers lost more ice in two years (2021–23) than in the thirty years between 1960 and 1990. In Antarctica, the world's largest iceberg (4,000 square kilometres) broke away and moved through the southern seas like a stately predator, a silent harbinger of rising sea levels. Ninety percent of marine life is at risk of extinction if greenhouse gases aren't curbed.

In his book *Anointed with Oil: How Christianity and Crude Made Modern America*, historian Darren Dochuk outlined the curious marriage between oil and evangelical Christianity. You would think that oil would be more closely aligned with the devil; it's found in his backyard, black as night, bringing obscene wealth as well as debauchery, corruption, and ruin. Venezuela's Juan Pablo Pérez Alphonzo, the founder of OPEC, noted, "We are drowning in the devil's excrement," and observed that it had, in fact, brought ruin to his country.

Oil began as a Christian endeavour in America, dominated by John D. Rockefeller and John Howard Pew, president and chair of Sun Oil (later Sunoco), the company founded by his father. Pew started the J. Howard Pew Freedom Trust "to promote the recognition of the interdependence of Christianity and freedom." Despite their religious affinities (Rockefeller was Baptist, Pew Presbyterian) Pew deplored what he saw as Rockefeller's liberal humanism, giving money to universities and scientific research, which he felt was fuelling America's slide into secularism. "We've allowed a lot of humanists to get control," he noted. To promote a "new evangelism" he donated heavily to Billy Graham's periodical *Christianity Today*, and helped lay the groundwork for the religious right that arrived decades later. Pew delivered sermons of his own, the most popular being "The Oil Industry: A Living Monument to the American System of Free Enterprise," and warned of the perils of Democrats, whom he characterized as "evil." Ahead of his time.

Pew's evangelical oil crusade has come full circle in Texas, where two Christian nationalist oil billionaires, Tim Dunn and Farris Wilks, have built a powerful political machine of think tanks, media organizations, action committees, and acolyte state politicians to create an authoritarian theocracy. Glenn Rogers, a Republican candidate who incurred the wrath of Dunn and Wilks, wrote, "History will prove that our current state government is the most corrupt ever and is 'bought' by a few radical dominionist billionaires seeking to destroy public education, privatize our public schools and create a theocracy." The term *dominionist* is from Genesis, where Man is given dominion over all living things.

Dunn's company, CrownQuest Operating, is one of the largest oil companies in Texas, and he believes his oil has existed for four thousand years, as Genesis suggests, not the 200 million that geologists affirm. Together, Dunn and Wilks have given $45 million to political action committees that support their causes, among them lower taxes for oil and bills that impede the growth of renewable energy. The ultimate goal is a "New Earth," which will come to pass "when heaven comes to earth and God dwells with his people as King." Dunn believes they are in a holy war that pits Christians against Marxists, an ideology created by Satan rather than Karl Marx, and includes not just Democrats but centrist Republicans. "It is becoming clear they want to kill us," Dunn has said. First, Texas will experience the "Great Awakening," then a grateful nation will follow.

Dunn was one of the ten biggest contributors to Trump's 2024 presidential campaign, partly because of his promise to empower Christians ("If I get in, you're going to be using that power at a level that you've never used before," Trump told a Nashville audience) and partly for Trump's support of oil. Evangelists viewed Trump's return to power as "a prophecy fulfilled." Dunn's rhetoric mirrored Trump's, labelling the left as communists and extremists, a threat to democracy. "The extremists want to de-industrialize America," Dunn said. "They want to live in huts around a campfire."

Sixty years before Trump courted oil and evangelicals, Barry Goldwater did the same, though he came to regret it. In 1964, Goldwater, a blunt, deeply conservative Southerner and, like Trump, an outsider, became the Republican candidate for president. He got the nomination in part due to the support of the oil industry and white evangelicals. Yet decades after his failed presidential bid (he lost by the largest margin in presidential history), Goldwater had misgivings about the creeping influence of white Christian nationalists. "Mark my word," he said in 1994, "if and when these preachers get control of the [Republican] party, and they're trying to do so, it's going to be a damn problem. Frankly, these people frighten me.

Politics and governing demand compromise. But these Christians believe they are acting in the name of God so they can't and won't compromise."

Thirty years later, white evangelicals are closer to gaining control of the Republican party, in part because they have learned to compromise. They endorsed Donald Trump, perhaps the most morally flawed politician in American history, a bar that has been set quite high. He has broken all but one of the commandments (number six, Thou shalt not murder), and the night is young, as they say. Christians embraced him largely because of his overturning of *Roe v. Wade* and his anti-LGBTQ stance, as well as his support for creationism, protecting prayer in schools, and posting the ten commandments in public buildings ("I LOVE THE TEN COMMANDMENTS IN PUBLIC SCHOOLS, PRIVATE SCHOOLS, AND MANY OTHER PLACES FOR THAT MATTER. READ IT—HOW CAN WE, AS A NATION, GO WRONG??? THIS MAY BE, IN FACT, THE FIRST MAJOR STEP IN THE REVIVAL OF RELIGION, WHICH IS DESPERATELY NEEDED, IN OUR COUNTRY," he posted on Truth Social in his trademark capitals). All of these are fundamental issues for the church, and in the process, white evangelicals have become as transactional as Trump himself, overlooking his sins to get what they want. Corinthians gives us a glimpse of how this ends. "For such men are false apostles, deceitful workmen, disguising themselves as apostles of Christ. And no wonder, for even Satan disguises himself as an angel of light. So it is no surprise if his servants, also, disguise themselves as servants of righteousness. Their end will correspond to their deeds."

In 2016, then–Oklahoma governor Mary Fallin declared October 13 Oilfield Prayer Day, a day for all

Christians to pray for the oil that God had created for them. In Texas, congregations gather over newly drilled wells, praying to hit oil, trying to keep the devil at bay. John Brown, a born-again Christian and owner of Texas-based Zion Oil & Gas, spent $130 million drilling dry holes in Israel, convinced by a passage in Genesis that there was oil there. Before taking office, Trump suggested Israel should bomb Iranian oil fields, though he wanted it done before he assumed power on January 20. Oil keeps circling back to the Holy Land.

Three thousand years before bitumen built Fort McMurray, it had built Babylon. There were large bitumen deposits in central Iraq, and, like Ernest Manning, King Nebuchadnezzar II prized it as "a symbol of progress and prosperity." He used it as mortar for buildings, to pave roads and seal ships. "And they had brick for stone, and bitumen for mortar," Genesis tells us. Babylon serves as a cautionary tale of wealth, power, self-indulgence, and the worship of false gods, falling finally, in Revelation 18, its fall heralding the Rapture. "Fallen, fallen is Babylon the great! It has become a dwelling place of demons, a haunt of every foul spirit…and the merchants of the earth have grown rich with the wealth of her wantonness…and the merchants of the earth weep and mourn for her, since no one buys their cargo anymore." A parable worth noting, especially if the term "merchants of the earth" is interpreted literally.

It only took an hour for mighty Babylon to fall, though the bible is notoriously elastic when it comes to time. Two thousand years later, American troops invaded Iraq in search of oil and set up a military base on the ruins of Babylon, one oil kingdom usurping another.

* * *

EVANGELICAL CHRISTIANS LIKE Tim Dunn believe that God alone can create a kingdom on earth. Instead, oil created a kingdom on earth, though it's sometimes difficult to tell them apart. Both are omnipresent, mysterious, sustaining, and glimpsed by a privileged few.

In the Book of Revelation, the seven years of fire and death finally end with Jesus descending, leading an army of angels to defeat Satan, and bringing peace for a thousand years. "And the beast was captured," Revelation 19 assures us, "and with it the false prophet [J.D. Vance?] who in its presence had worked the signs by which he deceived those who had received the mark of the beast and those who worshipped its image. These two were thrown alive into the lake of fire that burns with brimstone." Or at least into the fiery brimstone that is American politics.

For many of us, the End Days are less of a draw. We see the tribulations arrive, man-made rather than God-sent, worse every year. And we struggle to see the Rapture that follows. We think about our children and recall Eliot's lines from *The Waste Land*: "What are the roots that clutch, what branches grow / Out of this stony rubbish?"

Epilogue

LOOKING AT OUR current state, it's not hard to imagine a world that will look like an Ed Burtynsky photograph thirty years from now—toxic rivers, devastated landscapes, piles of plastic refuse stretching to the horizon. Trump has reversed environmental protections, just as he did in his first term. Pierre Poilievre wants to streamline environmental regulations that impede further fossil fuel production. Our annual Climate Change Conferences threaten to become a fossil fuel trade show.

Yet there is hope. Despite determined opposition against renewables in the oil fiefdoms of Alberta and Texas, both lead their countries in renewable energy production. Both have a long history with oil and harbour powerful mythologies, but eventually economic self-interest overcomes mythology. For most of the world, renewable power is now cheaper than fossil fuels. In 2022, the International Energy Agency announced that global investment in green energy had surpassed investment in fossil fuels for the first time. The IEA announced that solar photovoltaic power is "the cheapest electricity in history." Arab OPEC countries are aggressively installing solar energy for domestic use.

Global sales for EVs increased 35 percent in 2023 from the previous year; one in five new cars sold was electric. In 2024, Bloomberg reported that thirty-one countries had reached the tipping point in EV sales. The tipping point for new technologies is surprisingly low—a 5 percent adoption rate. Initially, sales are slow, whether it's colour TVs, smartwatches, or cars, but once they clear 5 percent, they tend to go mainstream and adoption rates can increase quickly. Thailand hit 5 percent EV sales in the first quarter of 2023; by the last quarter it was at 13 percent, fuelled in part by its first domestic EV factory, China's Great Wall Motor Co. In Norway, 90 percent of new cars are electric, and 60 percent in Sweden. Batteries continue to improve, helping ease the range anxiety that holds back some consumers (the Lucid Air has a range of 660 kilometres). The global network of public charging stations reached 8 million at the end of 2023.

* * *

AND PERHAPS THERE is hope in the unlikely form of China. While Trump was building walls, both literal and figurative, during his first term, China was building bridges with its Belt and Road program. Begun in 2013, it was designed to foster trade and connectivity throughout Asia, Africa, and Europe, and to challenge American hegemony on the international stage. China has signed agreements with 152 countries (among them forty-four sub-Saharan African countries and seventeen EU countries), providing loans (estimated at more than $1 trillion) and infrastructure, and gaining access to oil, gas, and minerals. Initially, some of the projects it financed were coal plants, but it has pivoted toward exporting green energy, where it is a global leader. Meanwhile, Trump is pursuing "energy dominance" with a fading, economically precarious fracking industry (more than two-thirds of America's oil and gas comes from fracking). His steadfast march into America's glorious past may end in 1973, with lineups at the pumps and riots in the streets.

Over the years, fracking has lost hundreds of billions of dollars. The head of commodities research at Goldman Sachs noted, "The industry, you know, it destroyed a lot of wealth, like 10 to 20 cents on every single dollar. I think the number is actually closer to 30 cents on every dollar." Why would the US prop up a money-losing, dated technology but be reluctant to put that money (hundreds of billions of dollars) into renewables? This is partly because the oil industry is politically entrenched in a way that renewables will never be. And oil remains a tool of geopolitics, a role that renewables are unlikely to fill, at least for America. And, finally, oil is powered by a compelling mythology, and no country relies on mythology as much as the US. The Cowboy, the American Dream, the American Way of Life—these are all powerful narratives that helped define a culture, and have infected other cultures, including our own.

Changing a narrative is a daunting task. Mythology, by design, is meant to be comforting. But at the heart of American myth is freedom and dominance, and neither will be found in oil.

* * *

THE MANHATTAN PROJECT was successful for a number of reasons, the first being that the West faced an imminent existential threat. Working in our favour, sadly, is the threat of climate change getting more imminent by the day. Building the atomic bomb under an impossible deadline required co-operation among government, academia, and the military, uncomfortable bedmates in normal times.

That kind of co-operation can be seen in Denmark's transition away from fossil fuels, which stems from private–public partnerships. A government paper noted, "While the public sector provides the ambitious long-term goals and stable framework conditions, the private sector supplies the innovation, solutions and investments needed to achieve the visions." The key here is a public sector that has stable, long-term goals, rather than one that dramatically changes direction with every election. Denmark is on track to lower emissions 70 percent below 1990 levels by 2030, and to be "a climate-neutral society" by 2050. The other Nordic countries, using roughly the same model, aren't far behind, leading the way in lowering emissions, EV adoption, and renewable energy.

* * *

WE CAN GET a glimpse of oil's short, violent life, and the road to a viable future, in Shetland. In 1969, oil was discovered in the North Sea. An article in *The Economist* stated, "For fifty years the Scottish National Party ran on oil. The discovery of deposits beneath the North Sea in 1969 transformed Britain's public finances and ignited a separatist movement." In the 1970s, the SNP slogan was

"It's Scotland's oil." Sullum Voe, one of Europe's largest oil terminals and the deepest port in the UK, was built in Shetland to accommodate the North Sea find. At its peak, it handled up to 1.5 million barrels per day. Shetland was no stranger to colonization—first Denmark, then Scotland, then Britain, and finally oil. Shetland flirted with independence from both Britain and Scotland, emboldened by its oil assets.

Shetland had to deal with the pressures on its infrastructure. It took 7,000 people to build the terminal, mostly from off the island, this in a population of 23,000. New roads, housing, and schools had to be built, the airport expanded. Shetland reaped the considerable financial benefits of oil and was faced with its environmental consequences. In 1993, the *Braer*, an oil tanker carrying 84,413 tonnes of Norwegian Gullfaks crude headed for Quebec, lost power off Shetland. It ran aground at Garth's Ness, a peninsula on the southwest coast, and began leaking oil almost immediately. The leakage continued for a week before the most intense North Atlantic cyclone on record arrived and the *Braer* broke apart, releasing its entire cargo into the sea, roughly twice the size of the *Exxon Valdez* spill. Along with the crude, there was 1,700 tonnes of heavy fuel oil and 125 tonnes of diesel fuel.

There were other environmental mishaps. In 1978, the oil tanker *Esso Bernicia* collided with concrete moorings at the terminal and spilled 1,174 tonnes of marine Bunker C fuel oil into the bay. There are no production standards for fuel oil; it is heavy, dirty, and more difficult to clean up than heavy crude. More than three thousand seabirds died. Material for new roads was quarried from a favourite landmark. Tankers coming into Sullom Voe were found to be deliberately dumping oil at sea as it saved them time and money. The North Sea, they argued, was already oily.

But the North Sea oil fields dwindled, and by 2017 the Sullum Voe Terminal was handling less than 100,000 bpd. That year, BP, which managed the terminal, announced it was turning over its duties to Enquest, "a specialist in end-of-life management of hydrocarbon resources." Sullum Voe would transition from being one of Europe's largest oil terminals to one of the largest green energy hubs of the future, housing green hydrogen production, offshore electrification, and carbon capture. It will store CO_2 from English emitters before it is shipped to storage reservoirs beneath the North Sea. This small island offers a microcosm of the short, complicated life of oil, and what the afterlife might bring. The green transition will be complex and expensive, and there will be inevitable delays and cost overruns and political gamesmanship and nostalgia for the days of oil, but Shetland offers a hopeful model.

The Oil Wars will always be with us. For more than a century the wars were among companies and nations trying to acquire and exploit territories that contained oil. But now oil is fighting its own war, facing both history and a green revolution, two battles it can't win.

Sources

BABYLON

Dochuk, Darren. *Anointed with Oil: How Christianity and Crude Made Modern America*. New York: Basic Books, 2019.

Nikiforuk, Andrew, Sheila Pratt, and Don Wanagas. *Running on Empty: Alberta After the Boom*. Edmonton: NeWest Press, 1987.

Gillmor, Don. "The Events Leading Up to Sir Norman Foster." *Walrus,* January/February 2008.

US Energy Information Administration

FIN DE SIÈCLE

Rich, Nathaniel. "Losing Earth: The Decade We Almost Stopped Climate Change." *New York Times Magazine,* August 1, 2018.

Bell, Alice. "Sixty Years of Climate Change Warnings: The Signs That Were Missed (and Ignored)." *Guardian,* July 5, 2021.

Sterba, James P. "Problems from Climate Changes Foreseen in a 1974 C.I.A. Report." New York Times, February 2, 1977.

"1982 Lodgepole Blowout and Its Legacy." Pembina Institute.

Thompson, Clive. "How 19th Century Scientists Predicted Global Warming." JSTOR Daily, December 17, 2019.

Franta, Benjamin. "COP 26 What Big Oil Knew About Climate Change, Starting in 1959." *Consortium News,* November 3, 2021.

MYTHOLOGY AND TRANSFORMATION

Conlin, Jonathan. "'Fouled by Oil'? Oil Diplomacy and the Lausanne Conference, 1914–1928." *International History Review,* August 20, 2024.

MacGregor, James G. *A History of Alberta*. Edmonton: Hurtig Publishers, 1981.

Reasons, Chuck, ed. *Stampede City: Power and Politics in the West*. Toronto: Between the Lines, 1984.

THE BATTLE BEGINS

Bell, Alice. "Sixty Years of Climate Change Warnings: The Signs That Were Missed (and Ignored)." *Guardian,* July 5, 2021.

Little, Amanda. "A Look Back at Reagan's Environmental Record." *Grist,* June 11, 2004.

Taft, Kevin. *Oil's Deep State: How the Petroleum Industry Undermines Democracy and Stops Action on Global Warming—in Alberta, and in Ottawa.* Toronto: James Lorimar & Company, 2017.

Taft, Kevin. *Shredding the Public Interest: Ralph Klein and 25 Years of One-Party Government.* Edmonton: University of Alberta Press and Parkland Institute, 1997.

Steward, Gillian. "Betting on Bitumen: Alberta's Energy Policies from Lougheed to Klein." Edmonton: Parkland Institute Report, June 2017.

Gillmor, Don. "The People's Choice." *Saturday Night,* August 1989.

THROUGH THE LOOKING-GLASS

Union of Concerned Scientists. "Climate Change Research Distorted and Suppressed." June 30, 2005.

Keefe, Josh. "Koch Brothers Receive $400 Million in Subsidies, Want Government out of Thanksgiving." *International Business News,* November 30, 2017.

Ahmed, Nafeez. "Iraq Invasion Was About Oil." *Guardian,* March 20, 2014.

Union of Concerned Scientists. "Climate Science vs. Fossil Fuel Fiction." Source material and background information, March 2015.

Mutitt, Greg. *Fuel on the Fire: Oil and Politics in Occupied Iraq.* New York: New Press, 2012.

Revkin, Andrew C., and Katherine Seelye. "Report Leaves Out Data on Climate Change." *New York Times,* June 19, 2003.

Coll, Steve. *Private Empire: ExxonMobil and American Power.* New York: Penguin Press, 2013.

Meyer, Robinson. "Trump's Obsession with Oil Could Destroy America's Auto Industry." *New York Times,* September 11, 2024.

Friedman, Lisa, Coral Davenport, Jonathan Swan, and Maggie Haberman. "At a Dinner, Trump Assailed Climate Rules and Asked $1 Billion from Big Oil." *New York Times,* May 9, 2024.

Taft, Kevin. *Oil's Deep State: How the Petroleum Industry Undermines Democracy and Stops Action on Global Warming—in Alberta, and in Ottawa.* Toronto: James Lorimer & Company, 2017.

SHIFTING SANDS

Gillmor, Don. "Shifting Sands." *Walrus,* April 2005.

Sekera, June, and Neva Goodwin. "Why the Oil Industry's Pivot to Carbon Capture and Storage—While It Keeps On Drilling—Isn't a Climate Solution." *Conversation,* November 23, 2021.

Misra, Siddharth. "Plummeting 'Energy Return on Investment' of Oil and the Impact on Global Energy Landscape." *Journal of Petroleum Technology,* March 20, 2023.

Wilt, James. "A Brief History of the Public Money Propping Up the Alberta Oilsands." *Narwhal,* May 16, 2018.

McNeill, Jodi, and Nina Lothian. "Review of Directive 085 Tailings Management Plans." Pembina Institute, March 13, 2017.
Smith, Charlie. "Report Shows 70 Percent of Canadian Oilsands Production Is Owned by Foreign Companies and Shareholders." *Georgia Straight,* May 11, 2020.
Kaplan, Lennie. "Canadian Upstream Oil Sector Supply Costs Continue to Decline." Calgary: Canadian Energy Centre, February 6, 2023.
Anderson, Mitchell. "IMF Pegs Canada's Fossil Fuel Subsidies at $34 Billion." *Tyee,* May 15, 2014.
Nickel, Rod, and Jeff Lewis. "Canada's Oil Patch Cuts Back Climate Efforts Under Pandemic." Reuters, June 14, 2020.
Cox, Sarah. "Canada's Oil and Gas Sector Received $18 Billion in Subsidies, Public Financing During Pandemic: Report." *Narwhal,* April 15, 2021.
Chung, Emily. "How Much Are Taxpayers Really Subsidizing Canada's Fossil Fuel Industry." CBC News, March 9, 2022.
Black, Simon. "IMF Fossil Fuel Subsidies Data: 2023 Update." International Monetary Fund, August 24, 2023.
Milstead, David. "Imperial Oil CEO Best-Paid in Canadian Energy Industry at $17.3 Million." *Globe and Mail,* April 20, 2023.
"Waiting to Launch 2022 Year-End Update." Pembina Institute, March 9, 2023.
Rosane, Olivia. "Fossil Fuel Consumption Subsidies Soared to Record Heights in 2022." World Economic Forum, February 21, 2023.
"Economic Impact of Oil and Gas." Canadian Association of Petroleum Producers.
Graney, Emma. "Cenovus CEO Jon McKenzie Labels Ottawa's Plans to Eliminate Oil and Gas Subsidies 'Political Rhetoric'." *Globe and Mail,* July 27, 2023.
Gillmor, Don. "The Oilsands' Third Act." *Arch,* Spring/Summer 2022.
Flanagan, Erin, and Jennifer Grant. "Losing Ground: Why the Problem of Oilsands Tailings Waste Keeps Growing." Pembina Institute, July 2013.
Bakx, Kyle. "Oilsands Giants Continue Work on Proposed $16.5B Carbon Capture Project, Despite Lingering Questions." CBC News, November 28, 2023.
"Capital Power Pulls Plug on Proposed $2.4B Genesee Carbon Capture and Storage Project." Canadian Press, May 1, 2024.

ORPHAN IN THE STORM

Weber, Bob. "Alberta Energy Minister Says Public Money Could Be Used in Well Cleanup." *Globe and Mail,* September 17, 2024.
Ensing, Chris. "Wheatley Explosion Could Be 'Tip of the Iceberg' in Ontario Given Number of Abandoned Wells: Expert." CBC News, Sepember 2, 2021.
Weber, Bob. "Alberta to Pilot Oil and Gas Royalty Breaks for Legally Required Well Cleanup." Canadian Press, 2023.
Singh, Inayat. "Alberta's Looming Multibillion-Dollar Orphan Wells Problem Prompts Auditor General Probe." CBC News, January 23, 2020.
Graney, Emma. "Catastrophe Looms Without Overhaul of Alberta's Inactive Oil and Gas Well Rules, Report Says." *Globe and Mail,* October 11, 2023.
Mercure, J.F., P. Salas, P. Vercoulen, G. Semieniuk, A. Lam, H. Pollitt, P.B. Holden, et al. "Reframing Incentives for Climate Policy Action." *Nature Energy,* November 4, 2021.

Heidenreich, Phil. "Alberta Renewable Energy Development Pause Affects $33B in Investment, 24,000 'Job-Years': Report." Global News, August 24, 2023.

Climenhaga, David. "Estimated 24,000 Jobs, $33B in Investments at Risk Because of Renewables Freeze." Pembina Institute, August 24, 2023.

Markusoff, Jason. "Assessing Danielle Smith's Latest Reasons for Pausing Alberta Wind and Solar, to the Letter." CBC News, August 15, 2023.

Alberta Electric System Operator. "AESO 2022 Annual Market Statistics." March 2023.

Schleussner, Carl-Friedrich, Marina Andrijevic, Jarmo Kiksha, Richard Heede, Joeri Rogelj, Sylvia Schmidt, and Holly Simpkin. "Carbon Majors' Trillion Dollar Damages." Climate Analytics, November 16, 2023.

Yewchuk, Drew, and Martin Olszinski. "The Liabilities Go Up and the Security Stays the Same: The Oilsands Mine Financial Security Program in 2024." University of Calgary Faculty of Law.

Graney, Emma. "Alberta Oil Sands Companies Won't Be Forced to Pay More to Cleanup Fund, Documents Show." *Globe and Mail*, October 3, 2024.

Meyer, Carl. "Oil and Gas Industry Lobbied Poilievre's Office Dozens of Times over a Year, Records Show." *Narwhal*, August 7, 2024.

Graney, Emma. "Catastrophe Looms Without Overhaul of Alberta's Inactive Oil and Gas Well Rules, Report Says." *Globe and Mail*, October 11, 2023.

Noel, Will. "Alberta's Renewable Energy Advantage." Blog, Pembina Institute, July 31, 2023.

Farrell, Maureen. "Larry Fink's Bet on Saudi Oil Money Is Also His Latest E.S.G. Woe." *New York Times*, August 3, 2023.

"Royal Bank the No 1 Financier of Fossil Fuel Development in the World, New Report Finds." Canadian Press, April 13, 2023.

Hussey, Ian. "Job Creation or Job Loss: Big Companies Use Tax Cut to Automate Away Jobs in the Oil Sands." Parkland Institute, October 3, 2022.

Groom, Nichola. "Millions of Abandoned Oil Wells Are Leaking Methane, a Climate Menace." *Science*, June 22, 2020.

Watts, Jonathan, Ashley Kirk, Niamh McIntyre, Pablo Gutierrez, and Niko Kommenda. "Half World's Fossil Fuel Assets Could Become Worthless by 2036 in Net Zero Transition." *Guardian*, November 4, 2021.

Wright, Ronald. *A Short History of Progress*. Toronto: Anansi, 2004.

Schwartz, Peter, and Doug Randall. "An Abrupt Climate Change Scenario and Its Implications for United States Security." Stanford University, October 2003.

OIL ADDICTION

Nakhoul, Samia, William Maclean, and Marwa Rashad. "Saudi Prince Unveils Sweeping Plans to End 'Addiction to Oil.'" Reuters, April 25, 2016.

Egan, Matt. "America and Saudi Arabia Are Locked in a Bitter Battle over Oil: The Stakes Are Massive." CNN Business, October 28, 2022.

MoheyEldin, Layla. "What Saudi Vision 2030 Means for the Future of Oil-Dependent Economies." *Glimpse from the Globe*, February 28, 2023.

Maass, Peter. *Crude World: The Violent Twilight of Oil*. New York: Vintage Books, 2010.

"Equatorial Guinea: Oil Wealth Squandered and Stolen." Human Rights Watch, June 15, 2017.

"Teodorin Obiang: French Court Fines Equatorial Guinea VP." BBC News, February 10, 2020.

ON OIL   139

DiChristopher, Tom. "Trump and GOP Killed an Energy Anti-Corruption Rule for No Good Reason, Advocates Say." CNBC, March 21, 2017.
Rushe, Dominic. "Donald Trump Lifts Anti-Corruption Rules in 'Gift to the American Oil Lobby'." *Guardian*, February 14, 2017.
Cadei, Emily. "Republicans Make It Easier to Keep Big Oil Payments to Foreign Governments a Secret." *Newsweek*, February 2, 2017.
Nikiforuk, Andrew. *The Energy of Slaves: Oil and the New Servitude*. Vancouver: Greystone Books and David Suzuki Foundation, 2012.
Smith, Jamie, and Aanu Adeoye. "Nigeria's Oil Sector Hit by Exodus of Foreign Companies." *Financial Times*, December 12, 2023.
Kakanov, Evgeny, Hansjorg Blochliger, and Lilas Demmou. "Resource Curse in Oil Exporting Countries." OECD Economics Department Working Papers No. 1511, August 16, 2021.

THE FRACKING REVOLUTION

Gruley, Bryan, Joe Carroll, and Asjylyn Loder. "The Incredible Rise and Final Hours of Fracking King Aubrey McClendon." Bloomberg, March 10, 2016.
McLean, Bethany. "How America's 'Most Reckless' Billionaire Created the Fracking Boom." *Guardian*, August 30, 2018.
"Fracking by the Numbers." Environment America, April 13, 2016.
Wilson, Kat. "Pennsylvania Watersheds at Risk: Drought and Fracking." Fractracker Alliance, September 8, 2023.
Brower, Derek, and Myles McCormick. "What the End of the US Shale Revolution Would Mean for the World." *Financial Times*, January 16, 2023.
Oberhaus, Daniel. "Fracking's Deadly Toll." *Harvard Magazine*, July–August 2022.
Weber, Bob. "Canada Leads G20 in Financing Fossil Fuels, Lags in Renewables, Report Says." Canadian Press, October 28, 2021.
Tamborrino, Kelsey, and Josh Siegal. "Big Winners from Biden's Climate Law: Republicans Who Voted Against It." *Politico*, January 23, 2023.

THE RAPTURE

The Bible, Revelation 8.
Banack, Clark. "God's Province: Evangelical Christianity, Political Thought and Conservatism in Alberta." Montreal and Kingston: McGill-Queen's University Press, 2016.
Kidd, Thomas. "Anointed with Oil: Evangelicals and the Petroleum Industry." Interview with Darren Dochuk, author of *Anointed with Oil: How Christianity and Crude Made Modern America*. Blog, *Gospel Coalition*, July 16, 2019.
Mantovani, Cecile. "Swiss Glaciers Lose 10% of Volume in Worst Two Years on Record, Watchdog Says." *Globe and Mail*, September 29, 2023.
Kofman, Ava. "How Two Billionaire Preachers Remade Texas Politics." *New York Times*, October 2, 2024.
Gold, Russell. "The Billionaire Bully Who Wants to Turn Texas into a Christian Theocracy." *Texas Monthly*, March 2024.
Phillips-Fein, Kim. "The Gospel of Oil." *Boston Review*, September 3, 2019.

DON GILLMOR is the author of *To the River*, which won the Governor General's Award for nonfiction. He is the author of four novels, *Breaking and Entering*, *Long Change*, *Mount Pleasant*, and *Kanata*, a two-volume history of Canada, *Canada: A People's History*, and nine books for children, two of which were nominated for the Governor General's Award. He was a senior editor at *The Walrus*, and his journalism has appeared in *Rolling Stone*, GQ, *Saturday Night*, *Toronto Life*, the *Globe and Mail*, and the *Toronto Star*. He has won twelve National Magazine Awards and numerous other honours. He lives in Toronto.